Nuclear Power

Nuclear Power

Deborah Black

Callisto Reference

www.callistoreference.com

Callisto Reference,
118-35 Queens Blvd., Suite 400,
Forest Hills, NY 11375, USA

Visit us on the World Wide Web at:
www.callistoreference.com

ISBN: 978-1-64116-597-6 (Hardback)

Cataloging-in-Publication Data

Nuclear power / Deborah Black.
p. cm.
Includes bibliographical references and index.
ISBN 978-1-64116-597-6
1. Nuclear energy. 2. Force and energy. 3. Nuclear physics. 4. Nuclear engineering.
5. Nuclear facilities. 6. Nuclear power plants. I. Black, Deborah.

TK9145 .N83 2022
333.792 4--dc23

Table of Contents

Preface

It is with great pleasure that I present this book. It has been carefully written after numerous discussions with my peers and other practitioners of the field. I would like to take this opportunity to thank my family and friends who have been extremely supporting at every step in my life.

The use of nuclear reactions that generate heat by releasing nuclear energy is known as nuclear power. This heat is often used in steam turbines to produce electricity in a nuclear power plant. Nuclear power can be obtained from nuclear fission, nuclear fusion and nuclear decay reactions. The nuclear fission of uranium and plutonium is responsible for producing the majority of electricity from nuclear power. The processes of nuclear decay are used in various applications such as radioisotope thermoelectric generators, medical imaging devices, etc. Nuclear power is considered to be one of the cleanest sources of energy in the world and has the lowest level of fatalities per unit of energy generated compared to other energy sources. This book elucidates the concepts and innovative models around prospective developments with respect to nuclear power. Some of the diverse topics covered herein address the varied types of nuclear plants that fall under this category. The extensive content of this book provides the readers with a thorough understanding of the subject.

The chapters below are organized to facilitate a comprehensive understanding of the subject:

Chapter - What is Nuclear Power?

Nuclear power refers to the utilization of nuclear reactions that release energy and generate heat. Such power can be obtained from nuclear fission, nuclear decay and nuclear fusion reactions. This chapter has been carefully written to provide an easy understanding of nuclear power.

Chapter - Significant Aspects of Nuclear Power

Some of the important aspects of nuclear power are nuclear physics, nuclear weapons, nuclear fuel, nuclear waste and reactor physics. Various types of nuclear fuels are used in nuclear power plants such as uranium-235 and plutonium-239. This chapter will briefly introduce all these significant aspects of nuclear power.

Chapter - Nuclear Reactions

Nuclear reaction refers to the change in the identity and traits of an atomic nucleus that are induced by bombarding it with an energetic particle such as alpha particle, neutron, proton, heavy ion, etc. Few significant areas of nuclear reactions are neutron activation, neutron capture, nuclear chain reaction, nuclear fission and nuclear fusion. All these key concepts related to nuclear reactions have been carefully analyzed in this chapter.

Chapter - Components of a Nuclear Power Plant

Some of the basic components of nuclear power plants are nuclear reactors, steam generators, reactor coolant pumps, cooling towers, spent fuel pool and containment building. The device which is used to initiate and control a self-sustained nuclear chain reaction is known as a nuclear reactor. The topics elaborated in this chapter will help in gaining a better perspective about these components of a nuclear power plant.

Deborah Black

1

What is Nuclear Power?

Nuclear power refers to the utilization of nuclear reactions that release energy and generate heat. Such power can be obtained from nuclear fission, nuclear decay and nuclear fusion reactions. This chapter has been carefully written to provide an easy understanding of nuclear power.

Nuclear power is the electricity generated by power plants that derive their heat from fission in a nuclear reactor. Except for the reactor, which plays the role of a boiler in a fossil-fuel power plant, a nuclear power plant is similar to a large coal-fired power plant, with pumps, valves, steam generators, turbines, electric generators, condensers, and associated equipment.

World Nuclear Power

Nuclear power provides almost 15 percent of the world's electricity. The first nuclear power plants, which were small demonstration facilities, were built in the 1960s. These prototypes provided "proof-of-concept" and laid the groundwork for the development of the higher-power reactors that followed.

The nuclear power industry went through a period of remarkable growth until about 1990, when the portion of electricity generated by nuclear power reached a high of 17 percent. That percentage remained stable through the 1990s and began to decline slowly around the turn of the 21st century, primarily because of the fact that total electricity generation grew faster than electricity from nuclear power while other sources of energy (particularly coal and natural gas) were able to grow more quickly to meet the rising demand. This trend appears likely to continue well into the 21st century. The Energy Information Administration (EIA), a statistical arm of the U.S. Department of Energy, has projected that world electricity generation between 2005 and 2035 will roughly double (from more than 15,000 terawatt-hours to 35,000 terawatt-hours) and that generation from all energy sources except petroleum will continue to grow.

In 2012 more than 400 nuclear reactors were in operation in 30 countries around the world, and more than 60 were under construction. The United States has the largest nuclear power industry, with more than 100 reactors; it is followed by France, which has more than 50. Of the top 15 electricity-producing countries in the world, all but two, Italy and Australia, utilize nuclear power to generate some of their electricity. The overwhelming majority of nuclear reactor generating capacity is concentrated in North America, Europe, and Asia. The early period of the nuclear power industry was dominated by North America (the United States and Canada), but in the 1980s that lead was overtaken by Europe. The EIA projects that Asia will have the largest nuclear capacity by 2035, mainly because of an ambitious building program in China.

A typical nuclear power plant has a generating capacity of approximately one gigawatt (GW; one billion watts) of electricity. At this capacity, a power plant that operates about 90 percent of the time (the U.S. industry average) will generate about eight terawatt-hours of electricity per year. The predominant types of power reactors are pressurized water reactors (PWRs) and boiling water reactors (BWRs), both of which are categorized as light water reactors (LWRs) because they use ordinary (light) water as a moderator and coolant. LWRs make up more than 80 percent of the world's nuclear reactors, and more than three-quarters of the LWRs are PWRs.

Issues affecting Nuclear Power

Countries may have a number of motives for deploying nuclear power plants, including a lack of indigenous energy resources, a desire for energy independence, and a goal to limit greenhouse gas emissions by using a carbon-free source of electricity. The benefits of applying nuclear power to these needs are substantial, but they are tempered by a number of issues that need to be considered, including the safety of nuclear reactors, their cost, the disposal of radioactive waste, and a potential for the nuclear fuel cycle to be diverted to the development of nuclear weapons.

Safety

The safety of nuclear reactors has become paramount since the Fukushima accident of 2011. Disaster included the need to (1) adopt risk-informed regulation, (2) strengthen management systems so that decisions made in the event of a severe accident are based on safety and not cost or political repercussions, (3) periodically assess new information on risks posed by natural hazards such as earthquakes and associated tsunamis, and (4) take steps to mitigate the possible consequences of a station blackout.

The four reactors involved in the Fukushima accident were first-generation BWRs designed in the 1960s. Newer Generation III designs, on the other hand, incorporate improved safety systems and rely more on so-called passive safety designs (i.e., directing cooling water by gravity rather than moving it by pumps) in order to keep the plants safe in the event of a severe accident or station blackout. For instance, in the Westinghouse AP1000 design, residual heat would be removed from the reactor by water circulating under the influence of gravity from reservoirs located inside the reactor's containment structure. Active and passive safety systems are incorporated into the European Pressurized Water Reactor (EPR) as well.

Traditionally, enhanced safety systems have resulted in higher construction costs, but passive safety designs, by requiring the installation of far fewer pumps, valves, and associated piping, may actually yield a cost saving.

Economics

A convenient economic measure used in the power industry is known as the levelized cost of electricity, or LCOE, which is the cost of generating one kilowatt-hour (kWh) of electricity averaged over the lifetime of the power plant. The LCOE is also known as the "busbar cost," as it represents the cost of the electricity up to the power plant's busbar, a conducting apparatus that links the plant's generators and other components to the distribution and transmission equipment that delivers the electricity to the consumer.

The busbar cost of a power plant is determined by 1) capital costs of construction, including finance costs, 2) fuel costs, 3) operation and maintenance (O&M) costs, and 4) decommissioning and waste-disposal costs. For nuclear power plants, busbar costs are dominated by capital costs, which can make up more than 70 percent of the LCOE. Fuel costs, on the other hand, are a relatively small factor in a nuclear plant's LCOE (less than 20 percent). As a result, the cost of electricity from a nuclear plant is very sensitive to construction costs and interest rates but relatively insensitive to the price of uranium. Indeed, the fuel costs for coal-fired plants tend to be substantially greater than those for nuclear plants. Even though fuel for a nuclear reactor has to be fabricated, the cost of nuclear fuel is substantially less than the cost of fossil fuel per kilowatt-hour of electricity generated. This fuel cost advantage is due to the enormous energy content of each unit of nuclear fuel compared to fossil fuel.

The O&M costs for nuclear plants tend to be higher than those for fossil-fuel plants because of the complexity of a nuclear plant and the regulatory issues that arise during the plant's operation. Costs for decommissioning and waste disposal are included in fees charged by electrical utilities. In the United States, nuclear-generated electricity was assessed a fee of $0.001 per kilowatt-hour to pay for a permanent repository of high-level nuclear waste. This seemingly modest fee yielded about $750 million per year for the Nuclear Waste Fund.

At the beginning of the 21st century, electricity from nuclear plants typically cost less than electricity from coal-fired plants, but this formula may not apply to the newer generation of nuclear power plants, given the sensitivity of busbar costs to construction costs and interest rates. Another major uncertainty is the possibility of carbon taxes or stricter regulations on carbon dioxide emissions. These measures would almost certainly raise the operating costs of coal plants and thus make nuclear power more competitive.

Radioactive-waste Disposal

Spent nuclear reactor fuel and the waste stream generated by fuel reprocessing contain radioactive materials and must be conditioned for permanent disposal. The amount of waste coming out of the nuclear fuel cycle is very small compared with the amount of waste generated by fossil fuel plants. However, nuclear waste is highly radioactive (hence its designation as high-level waste, or HLW), which makes it very dangerous to the public and the environment. Extreme care must be taken to ensure that it is stored safely and securely, preferably deep underground in permanent geologic repositories.

Despite years of research into the science and technology of geologic disposal, no permanent disposal site is in use anywhere in the world. In the last decades of the 20th century, the United States made preparations for constructing a repository for commercial HLW beneath Yucca Mountain,

Nevada, but by the turn of the 21st century, this facility had been delayed by legal challenges and political decisions. Pending construction of a long-term repository, U.S. utilities have been storing HLW in so-called dry casks aboveground. Some other countries using nuclear power, such as Finland, Sweden, and France, have made more progress and expect to have HLW repositories operational in the period 2020–25.

Proliferation

The claim has long been made that the development and expansion of commercial nuclear power led to nuclear weapons proliferation, because elements of the nuclear fuel cycle (including uranium enrichment and spent-fuel reprocessing) can also serve as pathways to weapons development. However, the history of nuclear weapons development does not support the notion of a necessary connection between weapons proliferation and commercial nuclear power.

The first pathway to proliferation, uranium enrichment, can lead to a nuclear weapon based on highly enriched uranium. It is considered relatively straightforward for a country to fabricate a weapon with highly enriched uranium, but the impediment historically has been the difficulty of the enrichment process. Since nuclear reactor fuel for LWRs is only slightly enriched (less than 5 percent of the fissile isotope uranium-235) and weapons need a minimum of 20 percent enriched uranium, commercial nuclear power is not a viable pathway to obtaining highly enriched uranium.

The second pathway to proliferation, reprocessing, results in the separation of plutonium from the highly radioactive spent fuel. The plutonium can then be used in a nuclear weapon. However, reprocessing is heavily guarded in those countries where it is conducted, making commercial reprocessing an unlikely pathway for proliferation. Also, it is considered more difficult to construct a weapon with plutonium versus highly enriched uranium.

More than 20 countries have developed nuclear power industries without building nuclear weapons. On the other hand, countries that have built and tested nuclear weapons have followed other paths than purchasing commercial nuclear reactors, reprocessing the spent fuel, and obtaining plutonium. Some have built facilities for the express purpose of enriching uranium; some have built plutonium production reactors; and some have surreptitiously diverted research reactors to the production of plutonium. All these pathways to nuclear proliferation have been more effective, less expensive, and easier to hide from prying eyes than the commercial nuclear power route. Nevertheless, nuclear proliferation remains a highly sensitive issue, and any country that wishes to launch a commercial nuclear power industry will necessarily draw the close attention of oversight bodies such as the International Atomic Energy Agency.

Advantages of Nuclear Power Generation

- Nuclear power generation does emit relatively low amounts of carbon dioxide (CO_2). The emissions of green house gases and therefore the contribution of nuclear power plants to global warming is therefore relatively little.
- This technology is readily available, it does not have to be developed first.
- It is possible to generate a high amount of electrical energy in one single plant.

Disadvantages of Nuclear Power Generation

- The problem of radioactive waste is still an unsolved one. The waste from nuclear energy is extremely dangerous and it has to be carefully looked after for several thousand years (10'000 years according to United States Environmental Protection Agency standards).
- High risks: Despite a generally high security standard, accidents can still happen. It is technically impossible to build a plant with 100% security. A small probability of failure will always last. The consequences of an accident would be absolutely devastating both for human being as for the nature. The more nuclear power plants (and nuclear waste storage shelters) are built, the higher is the probability of a disastrous failure somewhere in the world.
- Nuclear power plants as well as nuclear waste could be preferred targets for terrorist attacks. No atomic energy plant in the world could withstand an attack similar to 9/11 in New York. Such a terrorist act would have catastrophic effects for the whole world.
- During the operation of nuclear power plants, radioactive waste is produced, which in turn can be used for the production of nuclear weapons. In addition, the same know-how used to design nuclear power plants can to a certain extent be used to build nuclear weapons (nuclear proliferation).
- The energy source for nuclear energy is Uranium. Uranium is a scarce resource, its supply is estimated to last only for the next 30 to 60 years depending on the actual demand.
- The time frame needed for formalities, planning and building of a new nuclear power generation plant is in the range of 20 to 30 years in the western democracies.

2

Significant Aspects of Nuclear Power

Some of the important aspects of nuclear power are nuclear physics, nuclear weapons, nuclear fuel, nuclear waste and reactor physics. Various types of nuclear fuels are used in nuclear power plants such as uranium-235 and plutonium-239. This chapter will briefly introduce all these significant aspects of nuclear power.

Nuclear Physics

Nuclear physics is a field of physics that involves investigation of the building blocks and interactions of atomic nuclei. It includes studies of nuclear components such as protons and neutrons, forces such as the strong force (or strong interaction), and phenomena such as radioactive decay, nuclear fission, and nuclear fusion.

Nuclear power and nuclear weapons are the most commonly known applications of nuclear physics, but the research field is also the basis for a far wider range of less common applications, such as in medicine (nuclear medicine, magnetic resonance imaging), materials engineering (ion implantation), and archaeology(radiocarbon dating).

Related Fields

Nuclear physics is sometimes used synonymously with atomic physics, but physicists usually differentiate between the two. Atomic physics studies the combined system of the atomic nucleus and the arrangement of electrons around the nucleus.

Particle physics involves study of the elementary constituents of matter and radiation, and the interactions between them. Particle physics evolved out of nuclear physics and, for this reason, has been included under the same term in earlier times.

The discovery of the electron by J. J. Thomson was the first indication that the atom had internal structure. At the turn of the twentieth century, the accepted model of the atom was J. J. Thomson's "plum pudding" model, in which the atom was a large positively charged ball with small negatively charged electrons embedded inside it. By then, physicists had also discovered three types of radiation coming from atoms, which they named alpha, beta, and gamma radiation.

Experiments by Lise Meitner and Otto Hahn in 1911 and James Chadwick in 1914 showed that the beta decay spectrum was continuous rather than discrete. That is, electrons were ejected from the atom with a range of energies, rather than the discrete amounts of energies that were observed in gamma and alpha decays. This was a problem for nuclear physics at the time, because it indicated that energy was not conserved in these decays.

In 1905, Albert Einstein formulated the idea of mass–energy equivalence. While the work on radioactivity by Becquerel, Pierre and Marie Curie predated this, an explanation of the source of the energy of radioactivity had to await the discovery that the nucleus itself was composed of smaller constituents, the nucleons.

Discovery of the Nucleus by Rutherford's Team

In 1906, Ernest Rutherford published "Radiation of the α Particle from Radium in passing through Matter." Hans Geiger expanded on this work in a communication to the Royal Society with experiments he and Rutherford had done passing α particles through air, aluminum foil and gold leaf. More work was published in 1909, by Geiger and Ernest Marsden, and was greatly expanded in 1910 by Geiger. In 1911-12, Rutherford went before the Royal Society to explain the experiments and propound the new theory of the atomic nucleus as we now understand it.

The key experiment behind this announcement was performed in 1909, when Hans Geiger and Ernest Marsden, under Rutherford's supervision, fired alpha particles (helium nuclei) at a thin film of gold foil. The plum pudding model predicted that the alpha particles should come out of the foil with their trajectories being at most slightly bent. Their actual observations, however, were shocking: A few particles were scattered through large angles, with some bouncing completely backwards.

Those observations, upon analysis, led to the Rutherford model of the atom, in which the atom has a very small, dense nucleus containing most of its mass, and consisting of heavy positively charged particles with several embedded electrons that would (at least partially) balance out the charge (since the neutron was unknown). As an example of this model (which is not the modern one), nitrogen-14 was thought to consist of a nucleus with 14 protons and 7 electrons (21 total particles), and the nucleus was surrounded by 7 more orbiting electrons.

The Rutherford model worked quite well until studies of nuclear spin were carried out by Franco Rasetti at the California Institute of Technology in 1929. By 1925, it was known that protons and electrons had a spin of 1/2, and in the Rutherford model of nitrogen-14, 20 of the 21 particles should have paired up to cancel each other's spin, and the final odd particle should have left the nucleus with a spin of 1/2. Rasetti discovered, however, that nitrogen-14 has a spin of 1.

Discovery of Nuetron by Chadwick

In 1932, Chadwick realized that radiation that had been observed by Walther Bothe, Herbert L. Becker, and Irène and Frédéric Joliot-Curie was actually due to a neutral particle of about the same mass as the proton, that he called the neutron (following a suggestion about the need for such a particle, by Rutherford). That same year, Dmitri Ivanenko suggested that neutrons were in fact spin 1/2 particles and that the nucleus contained neutrons to explain the mass not due to protons, and that there were no electrons in the nucleus—only protons and neutrons. The neutron spin immediately solved the problem of the spin of nitrogen-14, as the one unpaired proton and one unpaired neutron in this model, each contribute a spin of 1/2 in the same direction, for a final total spin of 1.

With the discovery of the neutron, scientists could at last calculate what fraction of binding energy each nucleus had, from comparing the nuclear mass with that of the protons and neutrons that composed it. Differences between nuclear masses calculated in this way, and when nuclear reactions were measured, where found to agree with Einstein's calculation of the equivalence of mass and energy to high accuracy (within 1 percent, when calculated in 1934).

Yukawa's Meson Postulated to Bind Nuclei

In 1935, Hideki Yukawa proposed the first significant theory of the strong force to explain how neutrons and protons are held together in the nucleus. In the Yukawa interaction, a virtual particle, later called a meson, mediated a force between all nucleons, including protons and neutrons. This force explained why nuclei did not disintegrate under the influence of proton repulsion, and it also gave an explanation of why the attractive strong force had a more limited range than the electromagnetic repulsion between protons. Later, the discovery of the pi meson showed it to have the properties of Yukawa's particle.

With Yukawa's papers, the modern model of the atom was nearing completion. The center of the atom contains a tight ball of neutrons and protons, which is held together by the strong nuclear force, unless the nucleus is too large. Unstable nuclei may undergo alpha decay, when they emit an energetic helium nucleus, or beta decay, when they eject an electron (or positron). After one of these decays, the resultant nucleus may be left in an excited state, and in this case it decays to its ground state by emitting high energy photons (gamma decay).

The study of the strong and weak nuclear forces (the latter explained by Enrico Fermi via Fermi's interaction in 1934) led physicists to collide nuclei and electrons at ever higher energies. This research became the science of particle physics, the crown jewel of which is the standard model of particle physics, which unifies the strong, weak, and electromagnetic forces.

Modern Nuclear Physics

A heavy nucleus can contain hundreds of nucleons, which means that with some approximation it can be treated as a classical system, rather than a quantum-mechanical one. In the resulting liquid-drop model, the nucleus has an energy that arises partly from surface tension and partly from electrical repulsion of the protons. The liquid-drop model is able to reproduce many features of nuclei, including the general trend of binding energy with respect to mass number, as well as the phenomenon of nuclear fission.

Superimposed on this classical picture, however, are quantum mechanical effects, which can be described using the nuclear shell model, developed in large part by Maria Goeppert-Mayer. Nuclei with certain numbers of neutrons and protons (the magic numbers 2, 8, 20, 50, 82, 126, etc.) are particularly stable, because their shells are filled.

Other, more complicated models for the nucleus have also been proposed, such as the interacting boson model, in which pairs of neutrons and protons interact as bosons, analogously to Cooper pairs of electrons.

Much of current research in nuclear physics relates to the study of nuclei under extreme conditions, such as high spin and excitation energy. Nuclei may also have extreme shapes (similar to that of American footballs) or extreme neutron-to-proton ratios. Experimenters can create such nuclei using artificially induced fusion or nucleon transfer reactions, employing ion beams from a particle accelerator. Beams with even higher energies can be used to create nuclei at very high temperatures, and there are signs that these experiments have produced a phase transition from normal nuclear matter to a new state, the quark-gluon plasma, in which quarks mingle with one another, rather than being segregated in triplets as they are in neutrons and protons.

Modern Topics in Nuclear Physics

Nuclear Decay: Spontaneous Changes from One Nuclide to Another

There are 80 elements that have at least one stable isotope, and 250 such stable isotopes. However, there are thousands more well-characterized isotopes that are unstable. These radioisotopes may be unstable and decay along timescales ranging from fractions of a second to weeks, years, or even many millions of years.

If a nucleus has too few or too many neutrons it may be unstable, and will decay after some period of time. For example, in a process called beta decay, a nitrogen-16 atom (7 protons, 9 neutrons) is converted to an oxygen-16 atom (8 protons, 8 neutrons) within a few seconds of being created. In this decay, a neutron in the nitrogen nucleus is turned into a proton and an electron and antineutrino, by the weak nuclear force. The element is transmuted to another element in the process because, while it previously had seven protons (which makes it nitrogen), it now has eight (which makes it oxygen).

In alpha decay, the radioactive element decays by emitting a helium nucleus (2 protons and 2 neutrons), giving another element, plus helium-4. In many cases this process continues through several steps of this kind, including other types of decays, until a stable element is formed.

In gamma decay, a nucleus decays from an excited state into a lower energy state by emitting a gamma ray. The element is not changed in the process.

Other, more exotic decays, are also possible. For example, in internal conversion decay, the energy from an excited nucleus may be used to eject one of the inner orbital electrons from the atom. This process produces high speed electrons, but it is not beta decay, and (unlike beta decay) does not transmute one element to another.

Nuclear Fusion

When two light nuclei come into very close contact with each other, it is possible for the strong force to fuse the two together. It takes a great deal of energy to push the nuclei close enough together for the strong or nuclear forces to have an effect, so the process of nuclear fusion can take place only at very high temperatures or high densities. Once the nuclei are close enough together, the strong force overcomes their electromagnetic repulsion and squishes them into a new nucleus. A very large amount of energy is released when light nuclei fuse together because the binding energy per nucleon increases with mass number up until nickel-62.

Stars like the Sun are powered by the fusion of four protons into a helium nucleus, two positrons, and two neutrinos. The *uncontrolled*fusion of hydrogen into helium is known as thermonuclear runaway. Research to find an economically viable method of using energy from a *controlled* fusion reaction is currently being undertaken by various research establishments.

Nuclear Fission

For nuclei heavier than nickel-62, the binding energy per nucleon decreases with mass number. It is therefore possible for energy to be released if a heavy nucleus breaks apart into two lighter ones. This splitting of atomic nuclei is known as nuclear fission.

The process of alpha decay may be thought of as a special type of spontaneous nuclear fission. This process produces a highly asymmetrical fission because the four particles that make up the alpha particle are especially tightly bound to each other, making production of this nucleus in fission particularly likely.

For some of the heaviest nuclei that produce neutrons on fission, and which also easily absorb neutrons to initiate fission, a self-igniting type of neutron-initiated fission can be obtained, in a so-called chain reaction. Chain reactions were known in chemistry before physics, and in fact many familiar processes like fires and chemical explosions are chemical chain reactions.

The fission or "nuclear" chain-reaction, using fission-produced neutrons, is the source of energy for nuclear power plants and fission type nuclear bombs such as the two that the United States used against Hiroshima and Nagasaki to bring an end to World War II in the Pacific theater. Heavy nuclei such as uranium and thorium may undergo spontaneous fission, but they are much more likely to undergo alpha decay.

For a neutron-initiated chain-reaction to occur, there must be a critical mass of the element present in a certain space under certain conditions (these conditions slow and conserve neutrons for the reactions). There is one known example of a natural nuclear fission reactor, which was active in two regions of Oklo, Gabon, Africa, over 1.5 billion years ago. Measurements of natural neutrino emission have demonstrated that around half of the heat emanating from the Earth's core results from radioactive decay. However, it is not known if any of this results from fission chain-reactions.

Production of Heavy Elements

As the Universe cooled after the Big Bang, it eventually became possible for particles as we know them to exist. The most common particles created in the Big Bang that are still easily observable to us were protons (hydrogen) and electrons (in equal numbers). Some heavier elements were created as the protons collided with each other, but most of the heavy elements we see today were created inside of stars during a series of fusion stages, such as the proton-proton chain, the CNO cycle, and the triple-alpha process.

Progressively heavier elements are created during the evolution of a star. Since the binding energy per nucleon peaks around iron, energy is released only through fusion processes occurring below this point. Since the creation of heavier nuclei by fusion costs energy, nature resorts to the process of neutron capture. Neutrons (due to their lack of charge) are readily absorbed by a nucleus. The heavy elements are created by either a slow neutron capture process (the so-called *s* process) or by the rapid (or *r*) process. The *s* process occurs in thermally pulsing stars (called AGB, or asymptotic giant branch stars) and takes hundreds to thousands of years to reach the heaviest elements of lead and bismuth. The *r* process is thought to occur in supernova explosions because the conditions of high temperature, high neutron flux and ejected matter are present. These stellar conditions make the successive neutron captures very fast, involving very neutron-rich species, which then undergo beta decay to heavier elements, especially at the so-called waiting points that correspond to more stable nuclides with closed neutron shells (magic numbers). The *r* process duration is typically in the range of a few seconds.

Nuclear Weapon

A nuclear weapon (also called an atom bomb, nuke, atomic bomb, nuclear warhead, A-bomb, or nuclear bomb) is an explosive device that derives its destructive force from nuclear reactions, either fission (fission bomb) or from a combination of fission and fusion reactions (thermonuclear bomb). Both bomb types release large quantities of energy from relatively small amounts of matter. The first test of a fission ("atomic") bomb released an amount of energy approximately equal to 20,000 tons of TNT (84 TJ). The first thermonuclear ("hydrogen") bomb test released energy approximately equal to 10 million tons of TNT (42 PJ). A thermonuclear weapon weighing little more than 2,400 pounds (1,100 kg) can release energy equal to more than 1.2 million tons of TNT (5.0 PJ). A nuclear device no larger than traditional bombs can devastate an entire city by blast, fire, and radiation. Since they are weapons of mass destruction, the proliferation of nuclear weapons is a focus of international relations policy.

Nuclear weapons have been used twice in war, both times by the United States against Japan near the end of World War II. On August 6, 1945, the U.S. Army Air Forces detonated a uranium gun-type fission bomb nicknamed "Little Boy" over the Japanese city of Hiroshima; three days later, on August 9, the U.S. Army Air Forces detonated a plutonium implosion-type fission bomb nicknamed "Fat Man" over the Japanese city of Nagasaki. These bombings caused injuries that resulted in the deaths of approximately 200,000 civilians and military personnel. The ethics of these bombings and their role in Japan's surrender are subjects of debate.

Since the atomic bombings of Hiroshima and Nagasaki, nuclear weapons have been detonated over two thousand times for testing and demonstration. Only a few nations possess such weapons or are suspected of seeking them. The only countries known to have detonated nuclear weapons—and acknowledge possessing them—are (chronologically by date of first test) the United States, the Soviet Union (succeeded as a nuclear power by Russia), the United Kingdom, France, China, India, Pakistan, and North Korea. Israel is believed to possess nuclear weapons, though, in a policy of deliberate ambiguity, it does not acknowledge having them. Germany, Italy, Turkey, Belgium and the Netherlands are nuclear weapons sharing states. South Africa is the only country to have independently developed and then renounced and dismantled its nuclear weapons.

The Treaty on the Non-Proliferation of Nuclear Weapons aims to reduce the spread of nuclear weapons, but its effectiveness has been questioned, and political tensions remained high in the 1970s and 1980s. Modernisation of weapons continues to this day.

Types

In figure, the Trinity test of the Manhattan Project was the first detonation of a nuclear weapon, which led J. Robert Oppenheimer to recall verses from the Hindu scripture Bhagavad Gita: “If the radiance of a thousand suns were to burst at once into the sky, that would be like the splendor of the mighty one” and “I am become Death, the destroyer of worlds”.

There are two basic types of nuclear weapons: those that derive the majority of their energy from nuclear fission reactions alone, and those that use fission reactions to begin nuclear fusion reactions that produce a large amount of the total energy output.

Fission Weapons

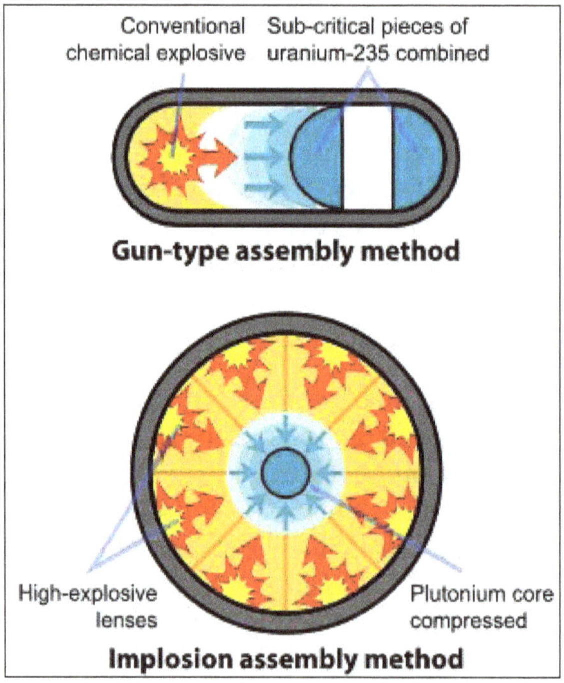

The two basic fission weapon designs.

All existing nuclear weapons derive some of their explosive energy from nuclear fission reactions. Weapons whose explosive output is exclusively from fission reactions are commonly referred to as atomic bombs or atom bombs (abbreviated as A-bombs). This has long been noted as something of a misnomer, as their energy comes from the nucleus of the atom, just as it does with fusion weapons.

In fission weapons, a mass of fissile material (enriched uranium or plutonium) is forced into supercriticality—allowing an exponential growth of nuclear chain reactions—either by shooting one piece of sub-critical material into another (the "gun" method) or by compression of a sub-critical sphere or cylinder of fissile material using chemically-fueled explosive lenses. The latter approach, the "implosion" method, is more sophisticated than the former.

A major challenge in all nuclear weapon designs is to ensure that a significant fraction of the fuel is consumed before the weapon destroys itself. The amount of energy released by fission bombs can range from the equivalent of just under a ton to upwards of 500,000 tons (500 kilotons) of TNT (4.2 to 2.1×10^{8} GJ).

All fission reactions generate fission products, the remains of the split atomic nuclei. Many fission products are either highly radioactive (but short-lived) or moderately radioactive (but long-lived), and as such, they are a serious form of radioactive contamination. Fission products are the principal radioactive component of nuclear fallout. Another source of radioactivity is the burst of free neutrons produced by the weapon. When they collide with other nuclei in surrounding material, the neutrons transmute those nuclei into other isotopes, altering their stability and making them radioactive.

The most commonly used fissile materials for nuclear weapons applications have been uranium-235 and plutonium-239. Less commonly used has been uranium-233. Neptunium-237 and some isotopes of americium may be usable for nuclear explosives as well, but it is not clear that this has ever been implemented, and their plausible use in nuclear weapons is a matter of dispute.

Fusion Weapons

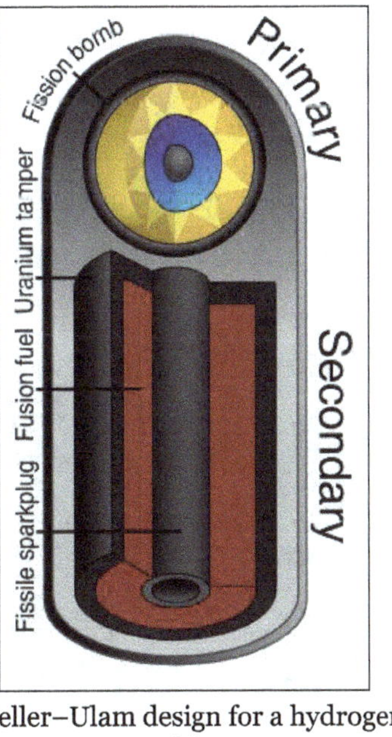

The basics of the Teller–Ulam design for a hydrogen bomb: a fission bomb uses radiation to compress and heat a separate section of fusion fuel.

The other basic type of nuclear weapon produces a large proportion of its energy in nuclear fusion reactions. Such fusion weapons are generally referred to as thermonuclear weapons or more colloquially as hydrogen bombs (abbreviated as H-bombs), as they rely on fusion reactions between isotopes of hydrogen (deuterium and tritium). All such weapons derive a significant portion of their energy from fission reactions used to "trigger" fusion reactions, and fusion reactions can themselves trigger additional fission reactions.

Only six countries—United States, Russia, United Kingdom, China, France, and India—have conducted thermonuclear weapon tests. (Whether India has detonated a "true" multi-staged thermonuclear weapon is controversial.) North Korea claims to have tested a fusion weapon as of January 2016, though this claim is disputed. Thermonuclear weapons are considered much more difficult to successfully design and execute than primitive fission weapons. Almost all of the nuclear weapons deployed today use the thermonuclear design because it is more efficient.

Thermonuclear bombs work by using the energy of a fission bomb to compress and heat fusion fuel. In the Teller-Ulam design, which accounts for all multi-megaton yield hydrogen bombs, this is accomplished by placing a fission bomb and fusion fuel (tritium, deuterium, or lithium deuteride) in proximity within a special, radiation-reflecting container. When the fission bomb is detonated, gamma rays and X-rays emitted first compress the fusion fuel, then heat it to thermonuclear temperatures. The ensuing fusion reaction creates enormous numbers of high-speed neutrons, which can then induce fission in materials not normally prone to it, such as depleted uranium. Each of these components is known as a "stage", with the fission bomb as the "primary" and the fusion capsule as the "secondary". In large, megaton-range hydrogen bombs, about half of the yield comes from the final fissioning of depleted uranium.

Virtually all thermonuclear weapons deployed today use the "two-stage" design described above, but it is possible to add additional fusion stages—each stage igniting a larger amount of fusion fuel in the next stage. This technique can be used to construct thermonuclear weapons of arbitrarily large yield, in contrast to fission bombs, which are limited in their explosive force. The largest nuclear weapon ever detonated, the Tsar Bomba of the USSR, which released an energy equivalent of over 50 megatons of TNT (210 PJ), was a three-stage weapon. Most thermonuclear weapons are considerably smaller than this, due to practical constraints from missile warhead space and weight requirements.

Edward Teller, often referred to as the "father of the hydrogen bomb".

Fusion reactions do not create fission products, and thus contribute far less to the creation of nuclear fallout than fission reactions, but because all thermonuclear weapons contain at least one fission stage, and many high-yield thermonuclear devices have a final fission stage, thermonuclear weapons can generate at least as much nuclear fallout as fission-only weapons.

Other Types

There are other types of nuclear weapons as well. For example, a boosted fission weapon is a fission bomb that increases its explosive yield through a small number of fusion reactions, but it is not a fusion bomb. In the boosted bomb, the neutrons produced by the fusion reactions serve primarily to increase the efficiency of the fission bomb. There are two types of boosted fission bomb: internally boosted, in which a deuterium-tritium mixture is injected into the bomb core, and externally boosted, in which concentric shells of lithium-deuteride and depleted uranium are layered on the outside of the fission bomb core.

Some nuclear weapons are designed for special purposes; a neutron bomb is a thermonuclear weapon that yields a relatively small explosion but a relatively large amount of neutron radiation; such a device could theoretically be used to cause massive casualties while leaving infrastructure mostly intact and creating a minimal amount of fallout. The detonation of any nuclear weapon is accompanied by a blast of neutron radiation. Surrounding a nuclear weapon with suitable materials (such as cobalt or gold) creates a weapon known as a salted bomb. This device can produce exceptionally large quantities of long-lived radioactive contamination. It has been conjectured that such a device could serve as a "doomsday weapon" because such a large quantity of radioactivities with half-lives of decades, lifted into the stratosphere where winds would distribute it around the globe, would make all life on the planet extinct.

In connection with the Strategic Defense Initiative, research into the nuclear pumped laser was conducted under the DOD program Project Excalibur but this did not result in a working weapon. The concept involves the tapping of the energy of an exploding nuclear bomb to power a single-shot laser which is directed at a distant target.

During the Starfish Prime high-altitude nuclear test in 1962, an unexpected effect was produced which is called a nuclear electromagnetic pulse. This is an intense flash of electromagnetic energy produced by a rain of high energy electrons which in turn are produced by a nuclear bomb's gamma rays. This flash of energy can permanently destroy or disrupt electronic equipment if insufficiently shielded. It has been proposed to use this effect to disable an enemy's military and civilian infrastructure as an adjunct to other nuclear or conventional military operations against that enemy. Because the effect is produced by high altitude nuclear detonations, it can produce damage to electronics over a wide, even continental, geographical area.

Research has been done into the possibility of pure fusion bombs: nuclear weapons that consist of fusion reactions without requiring a fission bomb to initiate them. Such a device might provide a simpler path to thermonuclear weapons than one that required development of fission weapons first, and pure fusion weapons would create significantly less nuclear fallout than other thermonuclear weapons, because they would not disperse fission products. In 1998, the United States Department of Energy divulged that the United States had, "made a substantial investment" in the past to develop pure fusion weapons, but that, "The U.S. does not have and is not developing a pure fusion weapon", and that, "No credible design for a pure fusion weapon resulted from the DOE investment".

Antimatter, which consists of particles resembling ordinary matter particles in most of their properties but having opposite electric charge, has been considered as a trigger mechanism for nuclear weapons. A major obstacle is the difficulty of producing antimatter in large enough quantities, and there is no evidence that it is feasible beyond the military domain. However, the U.S. Air Force funded studies of the physics of antimatter in the Cold War, and began considering its possible use in weapons, not just as a trigger, but as the explosive itself. A fourth generation nuclear weapon design is related to, and relies upon, the same principle as antimatter-catalyzed nuclear pulse propulsion.

Most variation in nuclear weapon design is for the purpose of achieving different yields for different situations, and in manipulating design elements to attempt to minimize weapon size.

Weapons Delivery

The first nuclear weapons were gravity bombs, such as this "Fat Man" weapon dropped on Nagasaki, Japan. They were large and could only be delivered by heavy bomber aircraft.

The system used to deliver a nuclear weapon to its target is an important factor affecting both nuclear weapon design and nuclear strategy. The design, development, and maintenance of delivery systems are among the most expensive parts of a nuclear weapons program; they account, for example, for 57% of the financial resources spent by the United States on nuclear weapons projects since 1940.

A demilitarized, commercial launch of the Russian Strategic Rocket Forces R-36 ICBM; also known by the NATO reporting name: SS-18 Satan. Upon its first fielding in the late 1960s, the SS-18 remains. the single highest throw weight missile delivery system ever built.

The simplest method for delivering a nuclear weapon is a gravity bomb dropped from aircraft; this was the method used by the United States against Japan. This method places few restrictions on the size of the weapon. It does, however, limit attack range, response time to an impending attack, and the number of weapons that a country can field at the same time. With miniaturization, nuclear bombs can be delivered by both strategic bombers and tactical fighter-bombers. This method is the primary means of nuclear weapons delivery; the majority of U.S. nuclear warheads, for example, are free-fall gravity bombs, namely the B61.

Montage of an inert test of a United States Trident SLBM (submarine launched ballistic missile), from submerged to the terminal, or re-entry phase, of the multiple independently targetable reentry vehicles.

Preferable from a strategic point of view is a nuclear weapon mounted on a missile, which can use a ballistic trajectory to deliver the warhead over the horizon. Although even short-range missiles allow for a faster and less vulnerable attack, the development of long-range intercontinental ballistic missiles (ICBMs) and submarine-launched ballistic missiles (SLBMs) has given some nations the ability to plausibly deliver missiles anywhere on the globe with a high likelihood of success.

More advanced systems, such as multiple independently targetable reentry vehicles (MIRVs), can launch multiple warheads at different targets from one missile, reducing the chance of a successful missile defense. Today, missiles are most common among systems designed for delivery of nuclear weapons. Making a warhead small enough to fit onto a missile, though, can be difficult.

Tactical weapons have involved the most variety of delivery types, including not only gravity bombs and missiles but also artillery shells, land mines, and nuclear depth charges and torpedoes for anti-submarine warfare. An atomic mortar has been tested by the United States. Small, two-man portable tactical weapons (somewhat misleadingly referred to as suitcase bombs), such as the Special Atomic Demolition Munition, have been developed, although the difficulty of combining sufficient yield with portability limits their military utility.

Nuclear Strategy

Nuclear warfare strategy is a set of policies that deal with preventing or fighting a nuclear war. The policy of trying to prevent an attack by a nuclear weapon from another country by threatening nuclear retaliation is known as the strategy of nuclear deterrence. The goal in deterrence is to always maintain a second strike capability (the ability of a country to respond to a nuclear attack with one of its own) and potentially to strive for first strike status (the ability to destroy an enemy's nuclear forces before they could retaliate). During the Cold War, policy and military theorists considered the sorts of policies that might prevent a nuclear attack, and they developed game theory models that could lead to stable deterrence conditions.

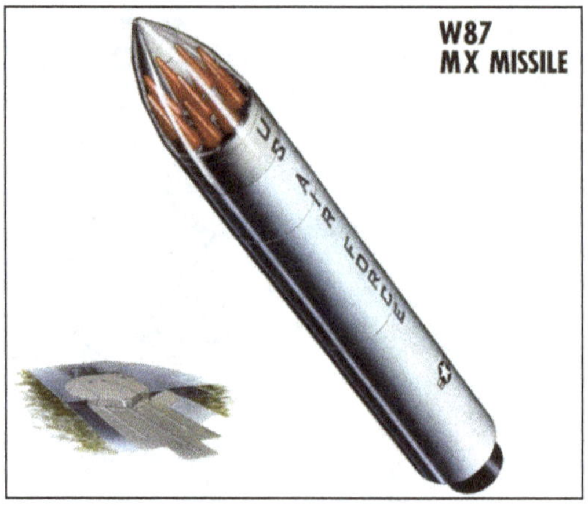

The now decommissioned United States' Peacekeeper missile was an ICBM developed to replace the Minuteman missile in the late 1980s. Each missile, like the heavier lift Russian SS-18 Satan, could contain up to ten nuclear warheads (shown in red), each of which could be aimed at a different target. A factor in the development of MIRVs was to make complete missile defense difficult for an enemy country.

Different forms of nuclear weapons delivery allow for different types of nuclear strategies. The goals of any strategy are generally to make it difficult for an enemy to launch a pre-emptive strike against the weapon system and difficult to defend against the delivery of the weapon during a potential conflict. This can mean keeping weapon locations hidden, such as deploying them on submarines or land mobile transporter erector launchers whose locations are difficult to track, or it can mean protecting weapons by burying them in hardened missile silo bunkers. Other components of nuclear strategies included using missile defenses to destroy the missiles before they land, or implementing civil defense measures using early-warning systems to evacuate citizens to safe areas before an attack.

Weapons designed to threaten large populations or to deter attacks are known as *strategic weapons*. Nuclear weapons for use on a battlefield in military situations are called *tactical weapons*.

Critics of nuclear war strategy often suggest that a nuclear war between two nations would result in mutual annihilation. From this point of view, the significance of nuclear weapons is to deter war because any nuclear war would escalate out of mutual distrust and fear, resulting in mutually assured destruction. This threat of national, if not global, destruction has been a strong motivation for anti-nuclear weapons activism.

Critics from the peace movement and within the military establishment have questioned the usefulness of such weapons in the current military climate. According to an advisory opinion issued by the International Court of Justice in 1996, the use of (or threat of use of) such weapons would generally be contrary to the rules of international law applicable in armed conflict, but the court did not reach an opinion as to whether or not the threat or use would be lawful in specific extreme circumstances such as if the survival of the state were at stake.

Another deterrence position is that nuclear proliferation can be desirable. In this case, it is argued that, unlike conventional weapons, nuclear weapons deter all-out war between states, and they succeeded in doing this during the Cold War between the U.S. and the Soviet Union. In the late 1950s and early 1960s, Gen. Pierre Marie Gallois of France, an adviser to Charles de Gaulle, argued in books like *The Balance of Terror: Strategy for the Nuclear Age* (1961) that mere possession of a nuclear arsenal was enough to ensure deterrence, and thus concluded that the spread of nuclear weapons could increase international stability. Some prominent neo-realist scholars, such as Kenneth Waltz and John Mearsheimer, have argued, along the lines of Gallois, that some forms of nuclear proliferation would decrease the likelihood of total war, especially in troubled regions of the world where there exists a single nuclear-weapon state. Aside from the public opinion that opposes proliferation in any form, there are two schools of thought on the matter: those, like Mearsheimer, who favored selective proliferation, and Waltz, who was somewhat more non-interventionist. Interest in proliferation and the stability-instability paradox that it generates continues to this day, with ongoing debate about indigenous Japanese and South Korean nuclear deterrent against North Korea.

The threat of potentially suicidal terrorists possessing nuclear weapons (a form of nuclear terrorism) complicates the decision process. The prospect of mutually assured destruction might not deter an enemy who expects to die in the confrontation. Further, if the initial act is from a stateless terrorist instead of a sovereign nation, there might not be a nation or specific target to retaliate against. It has been argued, especially after the September 11, 2001, attacks, that this complication calls for a new nuclear strategy, one that is distinct from that which gave relative stability during the Cold War. Since 1996, the United States has had a policy of allowing the targeting of its nuclear weapons at terrorists armed with weapons of mass destruction.

Robert Gallucci argues that although traditional deterrence is not an effective approach toward terrorist groups bent on causing a nuclear catastrophe, Gallucci believes that "the United States should instead consider a policy of expanded deterrence, which focuses not solely on the would-be nuclear terrorists but on those states that may deliberately transfer or inadvertently leak nuclear weapons and materials to them. By threatening retaliation against those states, the United States may be able to deter that which it cannot physically prevent".

Graham Allison makes a similar case, arguing that the key to expanded deterrence is coming up with ways of tracing nuclear material to the country that forged the fissile material. "After a nuclear bomb detonates, nuclear forensics cops would collect debris samples and send them to a laboratory for radiological analysis. By identifying unique attributes of the fissile material, including its impurities and contaminants, one could trace the path back to its origin." The process is analogous to identifying a criminal by fingerprints. "The goal would be twofold: first, to deter leaders of nuclear states from selling weapons to terrorists by holding them accountable for any use of their weapons; second, to give leaders every incentive to tightly secure their nuclear weapons and materials."

Governance, Control and Law

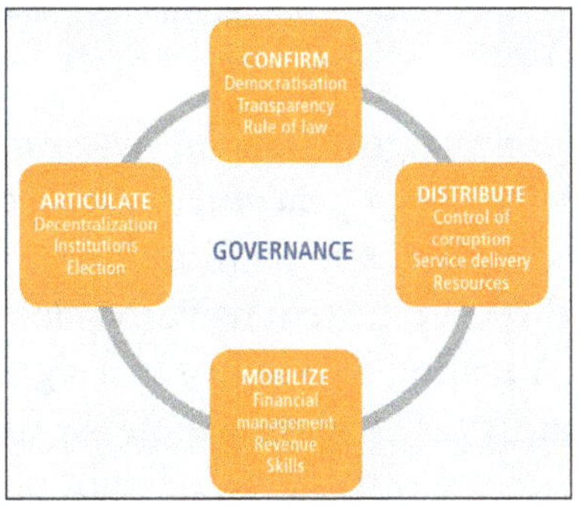

Because they are weapons of mass destruction, the proliferation and possible use of nuclear weapons are important issues in international relations and diplomacy. In most countries, the use of nuclear force can only be authorized by the head of government or head of state. Despite controls and regulations governing nuclear weapons, there is an inherent danger of "accidents, mistakes, false alarms, blackmail, theft, and sabotage".

In the late 1940s, lack of mutual trust prevented the United States and the Soviet Union from making progress on arms control agreements. The Russell–Einstein Manifesto was issued in London on July 9, 1955, by Bertrand Russell in the midst of the Cold War. It highlighted the dangers posed by nuclear weapons and called for world leaders to seek peaceful resolutions to international conflict. The signatories included eleven pre-eminent intellectuals and scientists, including Albert Einstein, who signed it just days before his death on April 18, 1955. A few days after the release, philanthropist Cyrus S. Eaton offered to sponsor a conference—called for in the manifesto—in Pugwash, Nova Scotia, Eaton's birthplace. This conference was to be the first of the Pugwash Conferences on Science and World Affairs, held in July 1957.

By the 1960s, steps were taken to limit both the proliferation of nuclear weapons to other countries and the environmental effects of nuclear testing. The Partial Nuclear Test Ban Treaty restricted all nuclear testing to underground nuclear testing, to prevent contamination from nuclear fallout, whereas the Treaty on the Non-Proliferation of Nuclear Weapons attempted to place restrictions on the types of activities signatories could participate in, with the goal of allowing the transference of non-military nuclear technology to member countries without fear of proliferation.

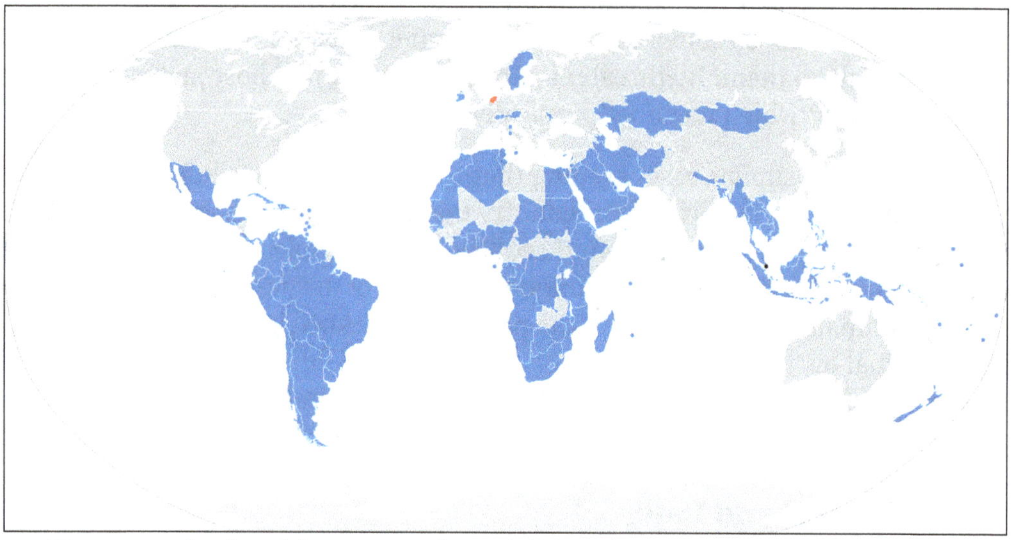

UN vote on adoption of the Treaty on the Prohibition of Nuclear Weapons on 7 July 2017.
Yes ■ No ■ Did not vote

In 1957, the International Atomic Energy Agency (IAEA) was established under the mandate of the United Nations to encourage development of peaceful applications of nuclear technology, provide international safeguards against its misuse, and facilitate the application of safety measures in its use. In 1996, many nations signed the Comprehensive Nuclear-Test-Ban Treaty, which prohibits all testing of nuclear weapons. A testing ban imposes a significant hindrance to nuclear arms development by any complying country. The Treaty requires the ratification by 44 specific states before it can go into force; as of 2012, the ratification of eight of these states is still required.

Additional treaties and agreements have governed nuclear weapons stockpiles between the countries with the two largest stockpiles, the United States and the Soviet Union, and later between the United States and Russia. These include treaties such as SALT II (never ratified), START I (expired), INF, START II (never ratified), SORT, and New START, as well as non-binding agreements such as SALT I and the Presidential Nuclear Initiatives of 1991. Even when they did not enter into force, these agreements helped limit and later reduce the numbers and types of nuclear weapons between the United States and the Soviet Union/Russia.

Nuclear weapons have also been opposed by agreements between countries. Many nations have been declared Nuclear-Weapon-Free Zones, areas where nuclear weapons production and deployment are prohibited, through the use of treaties. The Treaty of Tlatelolco prohibited any production or deployment of nuclear weapons in Latin America and the Caribbean, and the Treaty of Pelindaba prohibits nuclear weapons in many African countries. As recently as 2006 a Central Asian Nuclear Weapon Free Zone was established among the former Soviet republics of Central Asia prohibiting nuclear weapons.

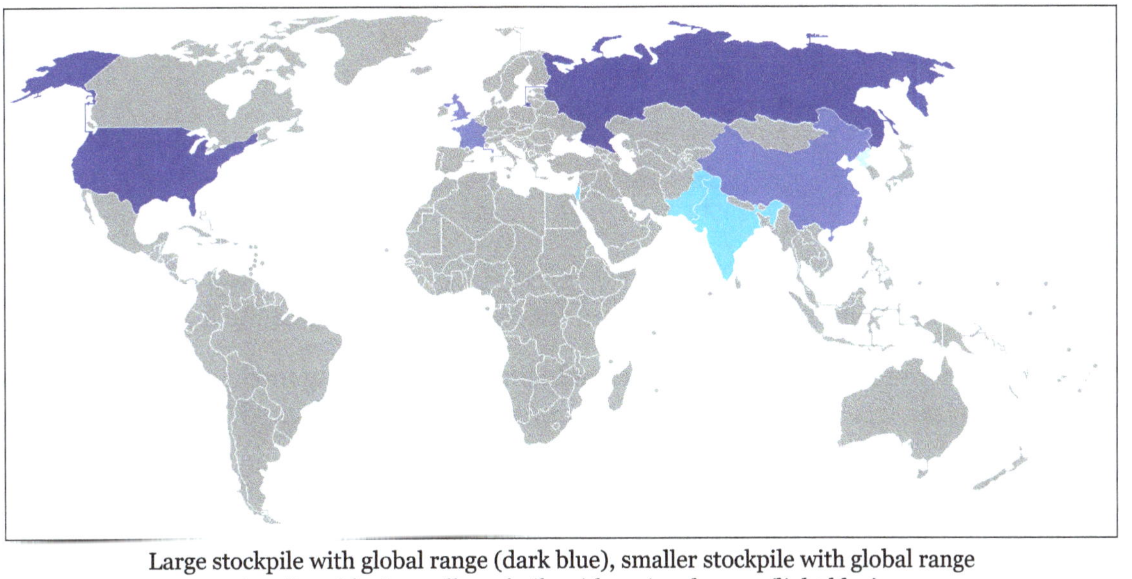

Large stockpile with global range (dark blue), smaller stockpile with global range (medium blue), small stockpile with regional range (light blue).

In 1996, the International Court of Justice, the highest court of the United Nations, issued an Advisory Opinion concerned with the "Legality of the Threat or Use of Nuclear Weapons". The court ruled that the use or threat of use of nuclear weapons would violate various articles of international law, including the Geneva Conventions, the Hague Conventions, the UN Charter, and the Universal Declaration of Human Rights. Given the unique, destructive characteristics of nuclear weapons, the International Committee of the Red Cross calls on States to ensure that these weapons are never used, irrespective of whether they consider them lawful or not.

Additionally, there have been other, specific actions meant to discourage countries from developing nuclear arms. In the wake of the tests by India and Pakistan in 1998, economic sanctions were (temporarily) levied against both countries, though neither were signatories with the Nuclear Non-Proliferation Treaty. One of the stated *casus belli* for the initiation of the 2003 Iraq War was an accusation by the United States that Iraq was actively pursuing nuclear arms (though this was soon discovered not to be the case as the program had been discontinued). In 1981, Israel had bombed a nuclear reactor being constructed in Osirak, Iraq, in what it called an attempt to halt Iraq's previous nuclear arms ambitions; in 2007, Israel bombed another reactor being constructed in Syria.

In 2013, Mark Diesendorf said that governments of France, India, North Korea, Pakistan, UK, and South Africa have used nuclear power and research reactors to assist nuclear weapons development or to contribute to their supplies of nuclear explosives from military reactors.

Disarmament

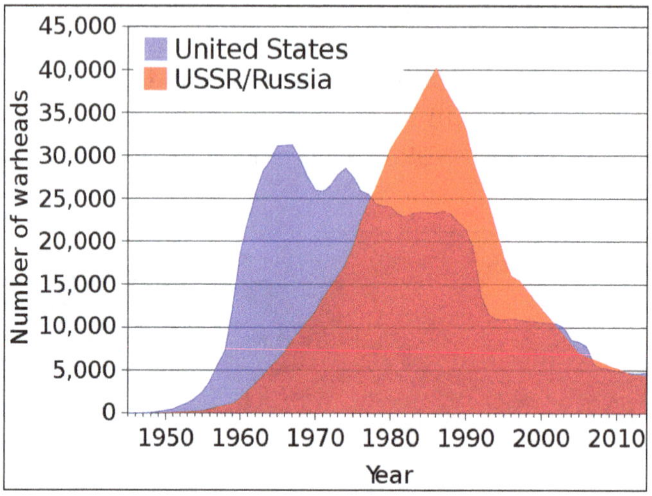

The USSR and United States nuclear weapon stockpiles throughout the Cold War until 2015, with a precipitous drop in total numbers following the end of the Cold War in 1991.

Nuclear disarmament refers to both the act of reducing or eliminating nuclear weapons and to the end state of a nuclear-free world, in which nuclear weapons are eliminated.

Beginning with the 1963 Partial Test Ban Treaty and continuing through the 1996 Comprehensive Test Ban Treaty, there have been many treaties to limit or reduce nuclear weapons testing and stockpiles. The 1968 Nuclear Non-Proliferation Treaty has as one of its explicit conditions that all signatories must "pursue negotiations in good faith" towards the long-term goal of "complete disarmament". The nuclear weapon states have largely treated that aspect of the agreement as "decorative" and without force.

Only one country—South Africa—has ever fully renounced nuclear weapons they had independently developed. The former Soviet republics of Belarus, Kazakhstan, and Ukraine returned Soviet nuclear arms stationed in their countries to Russia after the collapse of the USSR.

Proponents of nuclear disarmament say that it would lessen the probability of nuclear war, especially accidentally. Critics of nuclear disarmament say that it would undermine the present nuclear peace and deterrence and would lead to increased global instability. Various American elder statesmen, who were in office during the Cold War period, have been advocating the elimination of nuclear weapons. These officials include Henry Kissinger, George Shultz, Sam Nunn, and William Perry. In January 2010, Lawrence M. Krauss stated that "no issue carries more importance to the long-term health and security of humanity than the effort to reduce, and perhaps one day, rid the world of nuclear weapons".

Ukrainian workers use equipment provided by the U.S. Defense Threat Reduction Agency to dismantle a Soviet-era missile silo. After the end of the Cold War, Ukraine and the other non-Russian, post-Soviet republics relinquished Soviet nuclear stockpiles to Russia.

In the years after the end of the Cold War, there have been numerous campaigns to urge the abolition of nuclear weapons, such as that organized by the Global Zero movement, and the goal of a "world without nuclear weapons" was advocated by United States President Barack Obama in an April 2009 speech in Prague. A CNN poll from April 2010 indicated that the American public was nearly evenly split on the issue.

Some analysts have argued that nuclear weapons have made the world relatively safer, with peace through deterrence and through the stability–instability paradox, including in south Asia. Kenneth Waltz has argued that nuclear weapons have helped keep an uneasy peace, and further nuclear weapon proliferation might even help avoid the large scale conventional wars that were so common before their invention at the end of World War II. But former Secretary Henry Kissinger says there is a new danger, which cannot be addressed by deterrence: "The classical notion of deterrence was that there was some consequences before which aggressors and evildoers would recoil. In a world of suicide bombers, that calculation doesn't operate in any comparable way". George Shultz has said, "If you think of the people who are doing suicide attacks, and people like that get a nuclear weapon, they are almost by definition not deterrable".

As of early 2019, more than 90% of world's 13,865 nuclear weapons were owned by Russia and the United States.

United Nations

The UN Office for Disarmament Affairs (UNODA) is a department of the United Nations Secretariat established in January 1998 as part of the United Nations Secretary-General Kofi Annan's plan to reform the UN as presented in his report to the General Assembly in July 1997.

Its goal is to promote nuclear disarmament and non-proliferation and the strengthening of the disarmament regimes in respect to other weapons of mass destruction, chemical and biological weapons. It also promotes disarmament efforts in the area of conventional weapons, especially land mines and small arms, which are often the weapons of choice in contemporary conflicts.

Controversy

Ethics

Anti-nuclear weapons protest march..

Even before the first nuclear weapons had been developed, scientists involved with the Manhattan Project were divided over the use of the weapon. The role of the two atomic bombings of the country in Japan's surrender and the U.S.'s ethical justification for them has been the subject of scholarly and popular debate for decades. The question of whether nations should have nuclear weapons, or test them, has been continually and nearly universally controversial.

Notable Nuclear Weapons Accidents

- 21 August 1945: While conducting impromptu experiments on a third core (an alloy of plutonium and gallium) which had been prepared for atomic warfare at Los Alamos National Laboratory, physicist Harry Daghlian received a lethal dose of radiation. He died on 15 September 1945.
- 21 May 1946: While conducting further impromptu experiments on the third plutonium core at Los Alamos National Laboratory, physicist Louis Slotin received a lethal dose of radiation. He died on 30 May 1946. After these 2 incidents, the core was used to construct a bomb for use on the Nevada Test Range.
- February 13, 1950: A Convair B-36B crashed in northern British Columbia after jettisoning a Mark IV atomic bomb. This was the first such nuclear weapon loss in history.
- May 22, 1957: A 42,000-pound (19,000 kg) Mark-17 hydrogen bomb accidentally fell from a bomber near Albuquerque, New Mexico. The detonation of the device's conventional explosives destroyed it on impact and formed a crater 25 feet (7.6 m) in diameter on land owned by the University of New Mexico. According to a researcher at the Natural Resources Defense Council, it was one of the most powerful bombs made to date.
- June 7, 1960: The 1960 Fort Dix IM-99 accident destroyed a Boeing CIM-10 Bomarc nuclear missile and shelter and contaminated the BOMARC Missile Accident Site in New Jersey.
- January 24, 1961: The 1961 Goldsboro B-52 crash occurred near Goldsboro, North Carolina. A Boeing B-52 Stratofortress carrying two Mark 39 nuclear bombs broke up in mid-air, dropping its nuclear payload in the process.

- 1965 Philippine Sea A-4 crash, where a Skyhawk attack aircraft with a nuclear weapon fell into the sea. The pilot, the aircraft, and the B43 nuclear bomb were never recovered. It was not until 1989 that the Pentagon revealed the loss of the one-megaton bomb.
- January 17, 1966: The 1966 Palomares B-52 crash occurred when a B-52G bomber of the USAF collided with a KC-135 tanker during mid-air refuelling off the coast of Spain. The KC-135 was completely destroyed when its fuel load ignited, killing all four crew members. The B-52G broke apart, killing three of the seven crew members aboard. Of the four Mk28 type hydrogen bombs the B-52G carried, three were found on land near Almería, Spain. The non-nuclear explosives in two of the weapons detonated upon impact with the ground, resulting in the contamination of a 2-square-kilometer (490-acre) (0.78 square mile) area by radioactive plutonium. The fourth, which fell into the Mediterranean Sea, was recovered intact after a 2½-month-long search.
- January 21, 1968: The 1968 Thule Air Base B-52 crash involved a United States Air Force (USAF) B-52 bomber. The aircraft was carrying four hydrogen bombs when a cabin fire forced the crew to abandon the aircraft. Six crew members ejected safely, but one who did not have an ejection seat was killed while trying to bail out. The bomber crashed onto sea ice in Greenland, causing the nuclear payload to rupture and disperse, which resulted in widespread radioactive contamination. One of the bombs remains lost.
- September 18–19, 1980: The Damascus Accident, occurred in Damascus, Arkansas, where a Titan missile equipped with a nuclear warhead exploded. The accident was caused by a maintenance man who dropped a socket from a socket wrench down an 80-foot (24 m) shaft, puncturing a fuel tank on the rocket. Leaking fuel resulted in a hypergolic fuel explosion, jettisoning the W-53 warhead beyond the launch site.

Nuclear Testing and Fallout

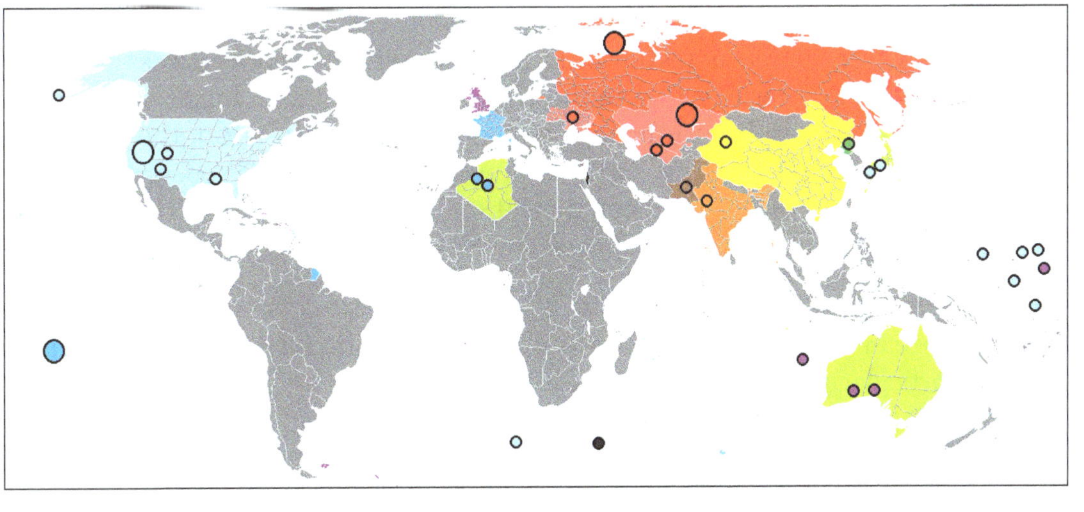

In figure, over 2,000 nuclear tests have been conducted in over a dozen different sites around the world. Red Russia/Soviet Union, blue France, light blue United States, violet Britain, yellow China, orange India, brown Pakistan, green North Korea and light green (territories exposed to nuclear bombs). The Black dot indicates the location of the Vela Incident.

This view of downtown Las Vegas shows a mushroom cloud in the background.
Scenes such as this were typical during the 1950s.

Over 500 atmospheric nuclear weapons tests were conducted at various sites around the world from 1945 to 1980. Radioactive fallout from nuclear weapons testing was first drawn to public attention in 1954 when the Castle Bravo hydrogen bomb test at the Pacific Proving Grounds contaminated the crew and catch of the Japanese fishing boat *Lucky Dragon*. One of the fishermen died in Japan seven months later, and the fear of contaminated tuna led to a temporary boycotting of the popular staple in Japan. The incident caused widespread concern around the world, especially regarding the effects of nuclear fallout and atmospheric nuclear testing, and "provided a decisive impetus for the emergence of the anti-nuclear weapons movement in many countries".

As public awareness and concern mounted over the possible health hazards associated with exposure to the nuclear fallout, various studies were done to assess the extent of the hazard. A Centers for Disease Control and Prevention/National Cancer Institute study claims that fallout from atmospheric nuclear tests would lead to perhaps 11,000 excess deaths among people alive during atmospheric testing in the United States from all forms of cancer, including leukemia, from 1951 to well into the 21st century. As of March 2009, the U.S. is the only nation that compensates nuclear test victims. Since the Radiation Exposure Compensation Act of 1990, more than $1.38 billion in compensation has been approved. The money is going to people who took part in the tests, notably at the Nevada Test Site, and to others exposed to the radiation.

In addition, leakage of byproducts of nuclear weapon production into groundwater has been an ongoing issue, particularly at the Hanford site.

Effects of Nuclear Explosions on Human Health

Some scientists estimate that a nuclear war with 100 Hiroshima-size nuclear explosions on cities could cost the lives of tens of millions of people from long term climatic effects alone. The climatology hypothesis is that *if* each city firestorms, a great deal of soot could be thrown up into the atmosphere which could blanket the earth, cutting out sunlight for years on end, causing the disruption of food chains, in what is termed a nuclear winter.

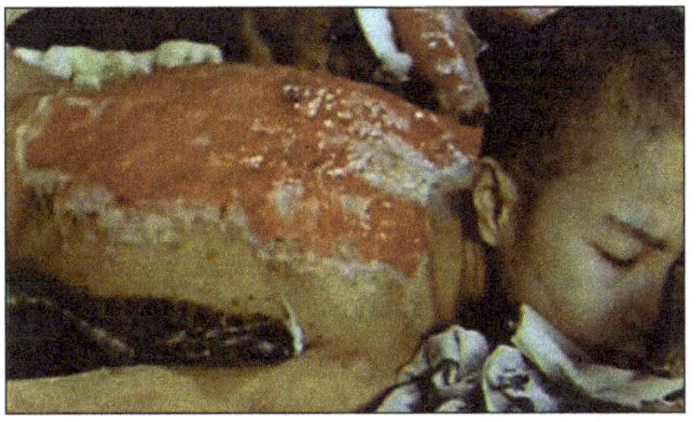

A photograph of Sumiteru Taniguchi's back injuries.

People near the Hiroshima explosion and who managed to survive the explosion subsequently suffered a variety of medical effects:

- Initial stage: The first 1–9 weeks, in which are the greatest number of deaths, with 90% due to thermal injury and blast effects and 10% due to super-lethal radiation exposure.
- Intermediate stage: From 10–12 weeks. The deaths in this period are from ionizing radiation in the median lethal range – LD50.
- Late period: Lasting from 13–20 weeks. This period has some improvement in survivors' condition.
- Delayed period: From 20+ weeks. Characterized by numerous complications, mostly related to healing of thermal and mechanical injuries, and if the individual was exposed to a few hundred to a thousand millisieverts of radiation, it is coupled with infertility, sub-fertility and blood disorders. Furthermore, ionizing radiation above a dose of around 50–100 millisievert exposure has been shown to statistically begin increasing one's chance of dying of cancer sometime in their lifetime over the normal unexposed rate of ~25%, in the long term, a heightened rate of cancer, proportional to the dose received, would begin to be observed after ~5+ years, with lesser problems such as eye cataracts and other more minor effects in other organs and tissue also being observed over the long term.

Fallout exposure – Depending on if further afield individuals shelter in place or evacuate perpendicular to the direction of the wind, and therefore avoid contact with the fallout plume, and stay there for the days and weeks after the nuclear explosion, their exposure to fallout, and therefore their total dose, will vary. With those who do shelter in place, and or evacuate, experiencing a total dose that would be negligible in comparison to someone who just went about their life as normal.

Staying indoors until after the most hazardous fallout isotope, I-131 decays away to 0.1% of its initial quantity after ten half lifes – which is represented by 80 days in I-131s case, would make the difference between likely contracting Thyroid cancer or escaping completely from this substance depending on the actions of the individual.

Public Opposition

Protest in Bonn against the nuclear arms race between the U.S./NATO and the Warsaw Pact, 1981.

Demonstration against nuclear testing in Lyon, France, in the 1980s.

Peace movements emerged in Japan and in 1954 they converged to form a unified "Japanese Council Against Atomic and Hydrogen Bombs". Japanese opposition to nuclear weapons tests in the Pacific Ocean was widespread, and "an estimated 35 million signatures were collected on petitions calling for bans on nuclear weapons".

In the United Kingdom, the first Aldermaston March organised by the Campaign for Nuclear Disarmament(CND) took place at Easter 1958, when, according to the CND, several thousand people marched for four days from Trafalgar Square, London, to the Atomic Weapons Research Establishment close to Aldermaston in Berkshire, England, to demonstrate their opposition to nuclear weapons. The Aldermaston marches continued into the late 1960s when tens of thousands of people took part in the four-day marches.

In 1959, a letter in the *Bulletin of the Atomic Scientists* was the start of a successful campaign to stop the Atomic Energy Commission dumping radioactive waste in the sea 19 kilometres from Boston. In 1962, Linus Pauling won the Nobel Peace Prize for his work to stop the atmospheric testing of nuclear weapons, and the "Ban the Bomb" movement spread.

In 1963, many countries ratified the Partial Test Ban Treaty prohibiting atmospheric nuclear testing. Radioactive fallout became less of an issue and the anti-nuclear weapons movement went into decline for some years. A resurgence of interest occurred amid European and American fears of nuclear war in the 1980s.

Costs and Technology Spin-offs

According to an audit by the Brookings Institution, between 1940 and 1996, the U.S. spent $9.3 trillion in present-day terms on nuclear weapons programs. 57 percent of which was spent on building nuclear weapons delivery systems. 6.3 percent of the total, $583 billion in present-day terms, was spent on environmental remediation and nuclear waste management, for example cleaning up the Hanford site, and 7 percent of the total, $653 billion was spent on making nuclear weapons themselves.

Non-weapons Uses

Peaceful nuclear explosions are nuclear explosions conducted for non-military purposes, such as activities related to economic development including the creation of canals. During the 1960s and 1970s, both the United States and the Soviet Union conducted a number of PNEs. Six of the explosions by the Soviet Union are considered to have been of an applied nature, not just tests.

Subsequently, the United States and the Soviet Union halted their programs. Definitions and limits are covered in the Peaceful Nuclear Explosions Treaty of 1976. The Comprehensive Nuclear-Test-Ban Treaty of 1996, once it enters into force, will prohibit all nuclear explosions, regardless of whether they are for peaceful purposes or not.

Reactor Physics

A knowledge of atomic and nuclear physics is essential to nuclear engineers, who deal with nuclear reactors. It should be noted that atomic and nuclear physics is very extensive branch of science. Nuclear reactor physics belongs to an applied physics as a particle physics or nuclear chemistry. These branches have common fundamentals. Atomic and nuclear physics describes fundamental particles (i.e. electrons, protons, neutrons), their structure, properties and behavior. These physical fundamentals consist of:

- Fundamental Particles.
- Atomic and Nuclear Structure.
- Mass and Energy.
- Radiation.
- Nuclear Stability.
- Radioactive Decay.
- Nuclear Reactions.
- Binding Energy.

Interaction of Radiation with Matter

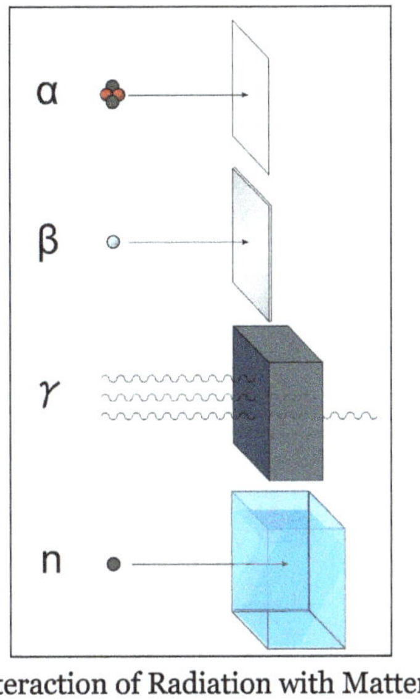

Interaction of Radiation with Matter.

The design of all nuclear reactors and other nuclear systems depends fundamentally on the way in which nuclear radiation interacts with matter. In nuclear reactor we can typically encounter with one of these types of radiation:

- Interactions of heavy charged particles.
- Interactions of beta radiation.
- Interactions of gamma rays.
- Interactions of neutrons.
- Reactor antineutrinos.

Neutron Diffusion Theory

Neutron Diffusion Theory is the branch of science that deals with the study and application of neutrons and its behavior within the nuclear core or in various environments. The Diffusion Theory provides theoretical basis for a neutron-physical computing of nuclear cores. The term theoretical has to be emphasized.

In order to design a nuclear reactor properly, it is necessary to be able to predict how the neutrons will be distributed throughout the system. This, in general, is a difficult problem, for the neutrons move in a reactor in complicated paths as a result of repeated collisions. To a first approximation the overall effect of these collisions is that the neutrons undergo a kind of diffusion in the reactor medium, much like the diffusion of one gas in another. This approximation allows solving such problem using the diffusion equation.

Nuclear Reactor Theory

The nuclear reactor theory is based on diffusion theory. The key term of the reactor theory is "criticality" of the reactor. Using the term "criticality" may seem counter-intuitive as a way to describe normalcy. The word often describes situations with potential for disaster. Nevertheless, in the context of nuclear power, "criticality" indicates that a reactor is operating safely and the neutron flux is stable at a given power level. In a critical reactor, there is a balance between the number of neutrons produced in fission and the number lost, either by absorbtion in the reactor or by leaking from its surface. The nuclear reactor theory solves the central problem in the design of a reactor, what is the calculation of the size and composition (nuclear fuel) of the system required to mantain this balance.

Reactor Kinetics – Zero Power Physics

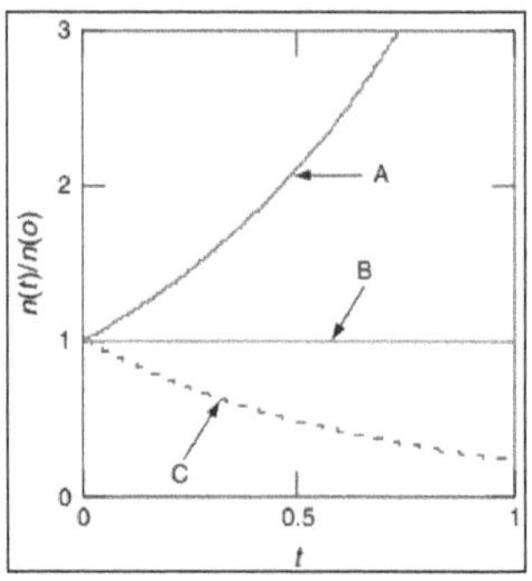

Reactor criticality. A – a supercritical state; B – a critical state; C – a subcritical state.

The nuclear reactor theory is concerned with critical nuclear reactors operating at constant power. But the reactor must not be always in the critical state. The reactor must be noncritical (supercritical or subcritical) in order to change the reactor power. The study of the behavior of neutron populations in a noncritical ractor is called reactor kinetics. This theory introduces a new unit – reactivity – of a nuclear reactor. The reactivity characterizes the deflection of reactor from the critical state.

It should be noted this theory can be used for a reactor at low power levels, hence "zero power criticality" (i.e. a reactor without reactivity feedbacks, 10E-8% – 1% of rated power). First periods of a reactor operation (reactor startup and power maneuvering at low power levels) are well described by this theory. Once the reactor core heats up (thermal power outweighs a decay power), reactivity feedbacks change the dynamics of the reactor.

Reactor Operation – Full Power Physics

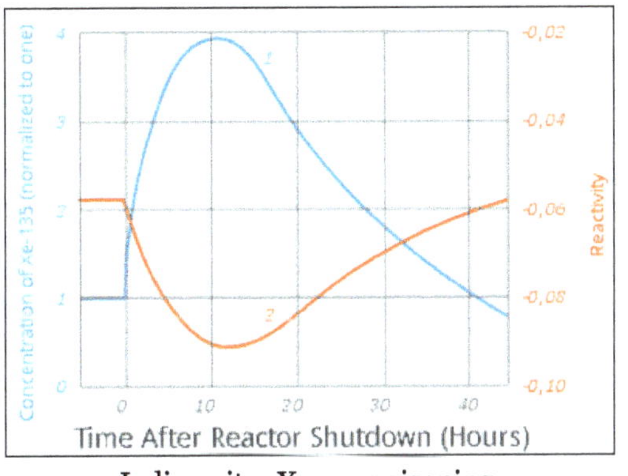

Iodine pit – Xenon poisoning.

This branch of nuclear engineering deals is very specific and dependent on a reactor type. PWRs have much in common with Boiling Water Reactors, but the operation of the reactors is not the same. On the other hand for example fast neutron reactors have completely different physics than PWRs and BWRs.

The degree of criticality at high power levels (~1% – 100% of rated power) is influenced by many variables. The most important are summarized below:

- Nuclear Fuel and Fuel Loading Pattern.
- Fuel Depletion.
- Neutron Leakage.
- Control Rods and Chemical Shim.
- Reactivity Feedbacks.
- Burnable Absorbers.
- Xenon and Samarium Poisoning.

Basic reactor operation physics includes these operation states:

- Reactor Startup ("going critical").

- Control of the Reactor and Power Maneuvering.
- Axial Offset Control.
- Xenon Oscillations.
- Reactor Shutdown.
- Emergency Shutdown – SCRAM.

Thermohydraulics of Nuclear Core

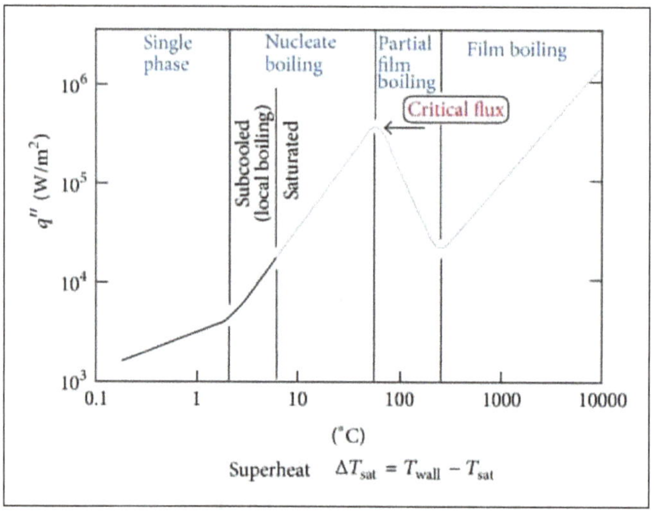

If the heat flux of a boiling system is higher than the critical heat flux then DNB.

Heat removal from nuclear reactors is also key issue of reactor physics. For a reactor to operate in a steady state, with an internal power distribution that is independent of time, all of the heat released in the system must be removed as fast as it is produced. This is accomplished by passing a coolant (water in case of PWRs or BWRs) through the core and through other regions where heat is generated.

The thermohydraulics deals with thermal behavior (heat generation, heat removal) and hydraulic behavior (Delta P, cross-flow issue, flow pattern) of the nuclear core. The nature and operation of this coolant system is one of the most important considerations on the design of a nuclear reactor. In the PWR reactors, there are phenomena which must be avoided. Namely DNB (Departure from Nucleate Boiling) and coolant saturation are very dangerous phenomena.

Behavior of Fuel Rod

Thermal and mechanical behavior of nuclear fuel constitute one of three key core design disciplines. Nuclear fuel is operated under very inhospitable conditions (thermal, radiation, mechanical) and must withstand more than normal conditions operation. For example temperatures in the centre of fuel pellets reach more than 1000 °C (1832 °F) accompanied by fission-gas releases. Behavior of nuclear fuel is associated with many phenomena. The most important are summarized below:

- Fuel swelling,
- Fuel densification,

- Thermal expansion,
- Fuel relocation,
- High Burnup Structure,
- Pellet-Cladding Interaction,
- Pellet-Coolant Interaction,
- Fission-Gas Releases.

Applied Core Physics

Core physics is the branch of science that deals with the study of behavior of specific nuclear core. The core physics provides methodological basis for a core design. Each nuclear core is due to a fuel reloading.

Full-scale physical computing of reactor core includes these types of computations:

- Core Design computing:
    - Fuel cross-section computing: Neutron cross-sections must be calculated for for fresh fuel assemblies usually in two-Dimensional models of the fuel lattice.
    - Loading Pattern search: The best arrangement of new fuel assemblies and used fuel assemblies is searched for in the core design from the viewpoints of safety and economy.
    - Fuel Cycle depletion: Behavior and key safety parameters of each proposed fuel cycle (power distribution, reactor kinetics parameters, control rods efficiency, etc.) must be analyzed for entire duration of the proposed cycle.
- Reload Safety Assessment:
    - Neutron-physical computing.
    - Thermal-hydraulic computing.
    - Fuel rod behavior computing.

Safety of Nuclear Reactors

'Safety' is the achievement of proper operating conditions, prevention of accidents and mitigation of accident consequences, resulting in protection of workers, the public and the environment from undue radiation hazards.

Nuclear safety is very broad branch of nuclear engineering. The three primary objectives of nuclear reactor safety systems as defined by the U.S. Nuclear Regulatory Commission are to shut down the reactor, maintain it in a shutdown condition and prevent the release of radioactive material. Appropriate redundant and diverse safety systems are installed to ensure safety of nuclear reactor.

Most of nuclear power plants introduce a 'defence-in-depth' approach to achieve maximum safety, this approach is constituted of multiple safety systems supplementing the natural features of the reactor core.

Nuclear Fuel

Nuclear fuel is the fuel that is used in a nuclear reactor to sustain a nuclear chain reaction. These fuels are fissile, and the most common nuclear fuels are the radioactive metals uranium-235 and plutonium-239. All processes involved in obtaining, refining, and using this fuel make up a cycle known as the nuclear fuel cycle.

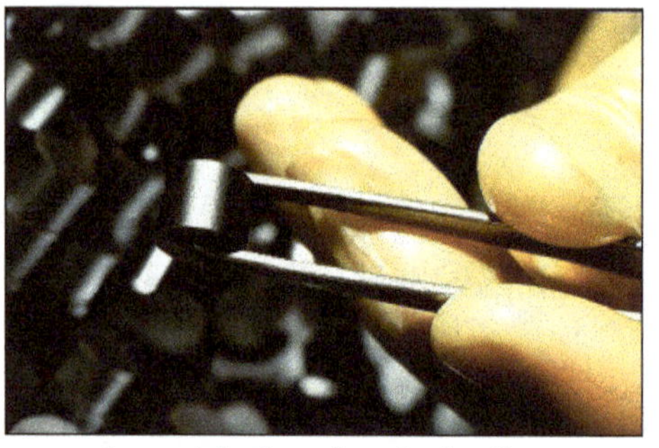

An enriched nuclear fuel pellet.

Uranium-235 is used as a fuel in different concentrations. Some reactors, such as the CANDU reactor, can use natural uranium with uranium-235 concentrations of only 0.7%, while other reactors require the uranium to be slightly enriched to levels of 3% to 5%. Plutonium-239 is produced and used in reactors (specifically fast breeder reactors) that contain significant amounts of uranium-238. It can also be recycled and used as a fuel in thermal reactors. Current research is being done to investigate how thorium-232 can be used as a fuel.

Production

Fuel fabrication plants are facilities that convert enriched uranium into fuel for nuclear reactors. For light water reactors, uranium is received from an enrichment plant in solid form. It is then converted into a gas and chemically converted into a uranium dioxide powder. This powder is then pressed into pellets and packed into fuel assemblies. A mixed oxide fuel can also be created when the uranium powder is packed along with plutonium oxide. The hazards present at fuel fabrication facilities—mainly chemical and radiological—are similar to the hazards at enrichment plants. These facilities generally pose a low risk to the public.

Use

When used in a reactor, the fuels used can have a variety of different forms a metal, an alloy, or some sort of oxide. Most nuclear reactors are fueled with a compound known as uranium dioxide. This uranium dioxide is put together in a fuel assembly and inserted into the nuclear reactor—where it can stay for several months or up to a few years. While in the reactor the fuel undergoes nuclear fission and releases energy. This released energy is used to generate electricity. Neutrons released during the fission process allow for a fission chain reaction to occur, allowing energy to be generated continually. The fuel is removed from the reactor after large amounts of the fuel—whether it is uranium-235 or plutonium-239—have undergone fission. The "used" nuclear fuel is known as spent or irradiated fuel. After use, the fuel must be cooled for a few years as it is extremely hot.

The spent fuel is placed in large, deep pools of water that act as a coolant and a radiation shield. The coolant property allows the water to remove the decay heat and the shielding abilities protect workers from the radioactivity of the fuel. After cooling, the fuel can be re-purposed or sent to storage depending on regulations.

Fuel Assembly

Nuclear reactors are powered by powdered uranium dioxide that has been compressed into small pellets, shown in figure. However, a power plant requires many of these pellets to run. Thus large numbers of these pellets are bundled into a fuel rod. A single uranium fuel pellet, only as big as a fingertip, contains as much energy as 481 cubic meters of natural gas, 807 kilograms of coal or 564 liters of oil. These rods are composed of numerous pellets of fissionable uranium fuel and can be several meters in length and about a centimeter in diameter. Then several of these rods, generally a dozen or more, are held together by strong metallic brackets in a fuel assembly. These rods are not bunched tightly together, rather there are several millimeters between each rod to allow coolant to flow between them. The tubes containing the pellets of uranium are generally composed of zirconium.

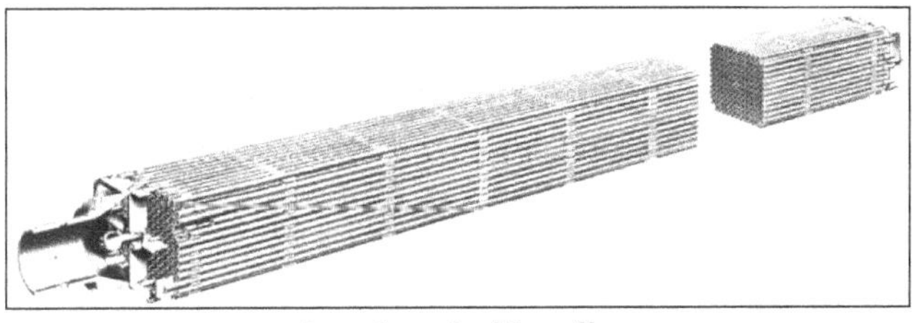

A nuclear fuel bundle.

Advantages and Disadvantages of Nuclear Fuels

Worldwide, there are extensive reserves of uranium left to mine. While nuclear fuel is not renewable, it is sustainable since there is so much of it. It will run out eventually, but not for centuries. Unlike fossil fuels, using nuclear fuels to produce energy does not directly produce carbon dioxide or sulfur dioxide. It should be mentioned that the processes of mining, transporting, and refining the fuel have carbon emissions associated with them, comparable to those of wind and solar power. Although the carbon footprint of using nuclear fuels is smaller, there are still disadvantages of using nuclear fuel. The waste, while a much lower volume must be handled very carefully because of its radioactivity. Nuclear fuels require far more complicated systems to extract their energy, which calls for greater regulation. These complex systems and regulation make for very long build times. In addition, public opinions on nuclear energy tend to be more negative than with other energy sources. The over-estimation of the dangers associated with releases of radioactive material is a significant issue, as large-scale nuclear incidents are rare.

Uranium-235

Uranium-235 (^{235}U) is an isotope of uranium making up about 0.72% of natural uranium. Unlike the predominant isotope uranium-238, it is fissile, i.e., it can sustain a fission chain reaction. It is the only fissile isotope that is primordial and found in relatively significant quantities in nature.

Uranium-235 has a half-life of 703.8 million years. It was discovered in 1935 by Arthur Jeffrey Dempster. Its fission cross section for slow thermal neutrons is about 584.994 barns. For fast neutrons it is on the order of 1 barn. Most but not all neutron absorptions result in fission; a minority result in neutron capture forming uranium-236.

Natural Decay Chain

$$ {}^{235}_{92}\mathrm{U} \xrightarrow[7.038\times10^{8}\ \mathrm{y}]{\alpha} {}^{231}_{90}\mathrm{Th} \xrightarrow[25.52\ \mathrm{h}]{\beta^-} {}^{231}_{91}\mathrm{Pa} \xrightarrow[3.276\times10^{4}\ \mathrm{y}]{\alpha} {}^{227}_{89}\mathrm{Ac} \left\{ \begin{array}{l} \xrightarrow[21.773\ \mathrm{y}]{98.62\%\beta^-} {}^{227}_{90}\mathrm{Th} \xrightarrow[18.718\ \mathrm{d}]{\alpha} \\ \xrightarrow[21.773\ \mathrm{y}]{1.38\%\alpha} {}^{223}_{87}\mathrm{Fr} \xrightarrow[21.8\ \mathrm{min}]{\beta^-} \end{array} \right\} {}^{223}_{88}\mathrm{Ra} \xrightarrow[11.434\ \mathrm{d}]{\alpha} {}^{219}_{86}\mathrm{Rn} $$

$$ {}^{219}_{86}\mathrm{Rn} \xrightarrow[3.96\ \mathrm{s}]{\alpha} {}^{215}_{84}\mathrm{Po} \xrightarrow[1.778\ \mathrm{ms}]{\alpha} {}^{211}_{82}\mathrm{Pb} \xrightarrow[36.1\ \mathrm{min}]{\beta^-} {}^{211}_{83}\mathrm{Bi} \left\{ \begin{array}{l} \xrightarrow[2.13\ \mathrm{min}]{99.73\%\alpha} {}^{207}_{81}\mathrm{Tl} \xrightarrow[4.77\ \mathrm{min}]{\beta^-} \\ \xrightarrow[2.13\ \mathrm{min}]{0.27\%\beta^-} {}^{211}_{84}\mathrm{Po} \xrightarrow[0.516\ \mathrm{s}]{\alpha} \end{array} \right\} {}^{207}_{82}\mathrm{Pb}_{(\mathrm{stable})} $$

Fission

The fission of one atom of uranium-235 releases 202.5 MeV = 3.24×10^{-11} J inside the reactor. That corresponds to 19.54 TJ/mol, or 83.14 TJ/kg. This is around 2.5 million times more than the energy released from burning coal. Another 8.8 MeV escapes the reactor as anti-neutrinos. When ${}^{235}_{92}\mathrm{U}$ nuclides are bombarded with neutrons, one of the many fission reactions that it can undergo is the following:

$$ {}^{1}_{0}\mathrm{n} + {}^{235}_{92}\mathrm{U} \rightarrow {}^{141}_{56}\mathrm{Ba} + {}^{92}_{36}\mathrm{Kr} + 3\,{}^{1}_{0}\mathrm{n} $$

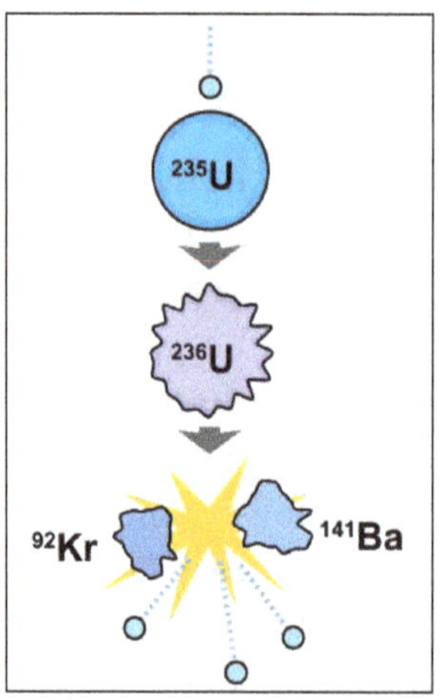

Nuclear fission seen with a uranium-235 nucleus.

Heavy water reactors, and some graphite moderated reactors, can use natural uranium, but light water reactors must use low enriched uranium because of the higher neutron absorption of light water. Uranium enrichment removes some of the uranium-238 and increases the proportion of uranium-235. Highly enriched uranium (HEU), which contains an even greater proportion of uranium-235, is sometimes used in nuclear submarines and nuclear weapons.

If at least one neutron from uranium-235 fission strikes another nucleus and causes it to fission, then the chain reaction will continue. If the reaction will sustain itself, it is said to be critical, and the mass of U-235 required to produce the critical condition is said to be a critical mass. A critical chain reaction can be achieved at low concentrations of U-235 if the neutrons from fission are moderated to lower their speed, since the probability for fission with slow neutrons is greater. A fission chain reaction produces intermediate mass fragments which are highly radioactive and produce further energy by their radioactive decay. Some of them produce neutrons, called delayed neutrons, which contribute to the fission chain reaction. The power output of nuclear reactors is adjusted by the location of control rods containing elements that strongly absorb neutrons, e.g., boron, cadmium, or hafnium, in the reactor core. In nuclear bombs, the reaction is uncontrolled and the large amount of energy released creates a nuclear explosion.

Nuclear Weapons

The Little Boy gun type atomic bomb dropped on Hiroshima on August 6, 1945 was made of highly enriched uranium with a large tamper. The nominal spherical critical mass for an untampered ^{235}U nuclear weapon is 56 kilograms (123 lb), a sphere 17.32 cm (6.8") in diameter. The material must be 85% or more of ^{235}U and is known as weapons grade uranium, though for a crude, inefficient weapon 20% is sufficient (called *weapon(s)-usable*). Even lower enrichment can be used, but then the required critical mass rapidly increases. Use of a large tamper, implosion geometries, trigger tubes, polonium triggers, tritium enhancement, and neutron reflectors can enable a more compact, economical weapon using one-fourth or less of the nominal critical mass, though this would likely only be possible in a country that already had extensive experience in engineering nuclear weapons. Most modern nuclear weapon designs use plutonium-239 as the fissile component of the primary stage, however HEU is often used in the secondary stage.

Source	Average energy released [MeV]
Instantaneously released energy	
Kinetic energy of fission fragments	169.1
Kinetic energy of prompt neutrons	4.8
Energy carried by prompt γ-rays	7.0
Energy from decaying fission products	
Energy of β^--particles	6.5
Energy of delayed γ-rays	6.3
Energy released when those prompt neutrons which don't (re)produce fission are captured	8.8
Total energy converted into heat in an operating thermal nuclear reactor	202.5
Energy of anti-neutrinos	8.8
Sum	211.3

Uses

Uranium-235 has many uses such as fuel for nuclear power plants, and nuclear weapons such as nuclear bombs.

Plutonium-239

Plutonium-239 is an isotope of plutonium. Plutonium-239 is the primary fissile isotope used for the production of nuclear weapons, although uranium-235 has also been used. Plutonium-239 is also one of the three main isotopes demonstrated usable as fuel in thermal spectrum nuclear reactors, along with uranium-235 and uranium-233. Plutonium-239 has a half-life of 24,110 years.

Nuclear Properties

The nuclear properties of plutonium-239, as well as the ability to produce large amounts of nearly pure Pu-239 more cheaply than highly enriched weapons-grade uranium-235, led to its use in nuclear weapons and nuclear power plants. The fissioning of an atom of uranium-235 in the reactor of a nuclear power plant produces two to three neutrons, and these neutrons can be absorbed by uranium-238 to produce plutonium-239 and other isotopes. Plutonium-239 can also absorb neutrons and fission along with the uranium-235 in a reactor.

Of all the common nuclear fuels, Pu-239 has the smallest critical mass. A spherical untamped critical mass is about 11 kg (24.2 lbs), 10.2 cm (4") in diameter. Using appropriate triggers, neutron reflectors, implosion geometry and tampers, this critical mass can be reduced by more than twofold. This optimization usually requires a large nuclear development organization supported by a sovereign nation.

The fission of one atom of Pu-239 generates 207.1 MeV = 3.318×10^{-11} J, i.e. 19.98 TJ/mol = 83.61 TJ/kg, or about 23,222,915 kilowatt hours/kg.

Radiation source (thermal fission of Pu-239)	Average energy released [MeV]
Kinetic energy of fission fragments	175.8
Kinetic energy of prompt neutrons	5.9
Energy carried by prompt γ-rays	7.8
Total instantaneous energy	189.5
Energy of β− particles	5.3
Energy of antineutrinos	7.1
Energy of delayed γ-rays	5.2
Total from decaying fission products	17.6
Energy released by radiative capture of prompt neutrons	11.5
Total heat released in a thermal-spectrum reactor (anti-neutrinos do not contribute)	211.5

Production

Plutonium is made from U-238. Pu-239 is normally created in nuclear reactors by transmutation of individual atoms of one of the isotopes of uranium present in the fuel rods. Occasionally, when an atom of U-238 is exposed to neutron radiation, its nucleus will capture a neutron, changing it to U-239. This happens more easily with lower kinetic energy (as U-238 fission activation is 6.6MeV). The U-239 then rapidly undergoes two β^- decays — an emission of an electron and an anti-neutrino ($\bar{\nu}_e$), leaving a proton — the first β^- decay transforming the U-239 into neptunium-239, and the second β^- decay transforming the Np-239 into Pu-239:

$$^{238}_{92}\mathrm{U} + {}^{1}_{0}\mathrm{n} \rightarrow {}^{239}_{92}\mathrm{U} \rightarrow \xrightarrow[23.5\ \mathrm{min}]{\beta^-} {}^{239}_{93}\mathrm{Np} \rightarrow \xrightarrow[2.356\ \mathrm{d}]{\beta^-} {}^{239}_{94}\mathrm{Pu}$$

Fission activity is relatively rare, so even after significant exposure, the Pu-239 is still mixed with a great deal of U-238 (and possibly other isotopes of uranium), oxygen, other components of the original material, and fission products. Only if the fuel has been exposed for a few days in the reactor, can the Pu-239 be chemically separated from the rest of the material to yield high-purity Pu-239 metal.

Pu-239 has a higher probability for fission than U-235 and a larger number of neutrons produced per fission event, so it has a smaller critical mass. Pure Pu-239 also has a reasonably low rate of neutron emission due to spontaneous fission (10 fission/s-kg), making it feasible to assemble a mass that is highly supercritical before a detonation chain reaction begins.

In practice, however, reactor-bred plutonium will invariably contain a certain amount of Pu-240 due to the tendency of Pu-239 to absorb an additional neutron during production. Pu-240 has a high rate of spontaneous fission events (415,000 fission/s-kg), making it an undesirable contaminant. As a result, plutonium containing a significant fraction of Pu-240 is not well-suited to use in nuclear weapons; it emits neutron radiation, making handling more difficult, and its presence can lead to a "fizzle" in which a small explosion occurs, destroying the weapon but not causing fission of a significant fraction of the fuel. (However, in modern nuclear weapons using neutron generators for initiation and fusion boosting to supply extra neutrons, fizzling is not an issue.) It is because of this limitation that plutonium-based weapons must be implosion-type, rather than gun-type. Moreover, Pu-239 and Pu-240 cannot be chemically distinguished, so expensive and difficult isotope separation would be necessary to separate them. Weapons-grade plutonium is defined as containing no more than 7% Pu-240; this is achieved by only exposing U-238 to neutron sources for short periods of time to minimize the Pu-240 produced.

Plutonium is classified according to the percentage of the contaminant plutonium-240 that it contains:

- Supergrade 2–3%,
- Weapons grade 3–7%,
- Fuel grade 7–18%,
- Reactor grade 18% or more.

A nuclear reactor that is used to produce plutonium for weapons therefore generally has a means for exposing U-238 to neutron radiation and for frequently replacing the irradiated U-238 with new U-238. A reactor running on unenriched or moderately enriched uranium contains a great deal of U-238. However, most commercial nuclear power reactor designs require the entire reactor to shut down, often for weeks, in order to change the fuel elements. They therefore produce plutonium in a mix of isotopes that is not well-suited to weapon construction. Such a reactor could have machinery added that would permit U-238 slugs to be placed near the core and changed frequently, or it could be shut down frequently, so proliferation is a concern; for this reason, the International Atomic Energy Agency inspects licensed reactors often. A few commercial power reactor designs, such as the *reaktor bolshoy moshchnosti kanalniy* (RBMK) and pressurized heavy water reactor (PHWR), do permit refueling without shutdowns, and they may pose a proliferation risk. (In fact, the RBMK was built by the Soviet Union during the Cold War, so despite their ostensibly peaceful purpose, it is likely that plutonium production was a design criterion.) By contrast, the Canadian CANDU heavy-water moderated natural-uranium fueled reactor can also be refueled while operating, but it normally consumes most of the Pu-239 it produces *in situ;* thus, it is not only inherently less proliferative than most reactors, but can even be operated as an "actinide incinerator." The American IFR (Integral Fast Reactor) can also be operated in an "incineration mode," having some advantages in not building up the Pu-242 isotope or the long-lived actinides, either of which cannot be easily burned except in a fast reactor. Also IFR fuel has a high proportion of burnable isotopes, while in CANDU an inert material is needed to dilute the fuel; this means the IFR can burn a higher fraction of its fuel before needing reprocessing. Most plutonium is produced in research reactors or plutonium production reactors called breeder reactors because they produce more plutonium than they consume fuel; in principle, such reactors make extremely efficient use of natural uranium. In practice, their construction and operation is sufficiently difficult that they are generally only used to produce plutonium. Breeder reactors are generally (but not always) fast reactors, since fast neutrons are somewhat more efficient at plutonium production.

Plutonium-239 is more frequently used in nuclear weapons than uranium-235, as it is easier to obtain in a quantity of critical mass. Both Plutonium-239 and Uranium-235 are obtained from Natural uranium, which primarily consists of uranium-238 but contains traces of other isotopes of uranium such as Uranium-235. The process of enriching uranium, i.e. increasing the ratio of U-235 to U-238 to weapons grade, is generally a more lengthy and costly process than the production of plutonium 239 from U-238 and subsequent reprocessing.

Supergrade Plutonium

The "supergrade" fission fuel, which has less radioactivity, is used in the primary stage of US Navy nuclear weapons in place of the conventional plutonium used in the Air Force's versions. "Supergrade" is industry parlance for plutonium alloy bearing an exceptionally high fraction of Pu-239 (>95%), leaving a very low amount of Pu-240 which is a high spontaneous fission isotope . Such plutonium is produced from fuel rods that have been irradiated a very short time as measured in MW-day/ton burnup. Such low irradiation times limit the amount of additional neutron capture and therefore buildup of alternate isotope products such as Pu-240 in the rod, and also by consequence is considerably more expensive to produce, needing far more rods irradiated and processed for a given amount of plutonium.

Plutonium-240, in addition to being a neutron emitter after fission, is a gamma emitter in that process as well, and so is responsible for a large fraction of the radiation from stored nuclear weapons. Whether out on patrol or in port, submarine crew members routinely live and work in very close proximity to nuclear weapons stored in torpedo rooms and missile tubes, unlike Air Force missiles where exposures are relatively brief. The need to reduce radiation exposure justifies the additional costs of the premium supergrade alloy used on many naval nuclear weapons. Supergrade plutonium is used in W80 warheads.

Nuclear Power Reactors

In any operating nuclear reactor containing U-238, some plutonium-239 will accumulate in the nuclear fuel. Unlike reactors used to produce weapons-grade plutonium, commercial nuclear power reactors typically operate at a high burnup that allows a significant amount of plutonium to build up in irradiated reactor fuel. Plutonium-239 will be present both in the reactor core during operation and in spent nuclear fuel that has been removed from the reactor at the end of the fuel assembly's service life (typically several years). Spent nuclear fuel commonly contains about 0.8% plutonium-239.

Plutonium-239 present in reactor fuel can absorb neutrons and fission just as uranium-235 can. Since plutonium-239 is constantly being created in the reactor core during operation, the use of plutonium-239 as nuclear fuel in power plants can occur without reprocessing of spent fuel; the plutonium-239 is fissioned in the same fuel rods in which it is produced. Fissioning of plutonium-239 provides about one-third of the total energy produced in a typical commercial nuclear power plant. Reactor fuel would accumulate much more than 0.8% plutonium-239 during its service life if some plutonium-239 were not constantly being "burned off" by fissioning.

A small percentage of plutonium-239 can be deliberately added to fresh nuclear fuel. Such fuel is called MOX (mixed oxide) fuel, as it contains a mixture of uranium oxide (UO_2) and plutonium oxide (PuO_2). The addition of plutonium-239 reduces the need to enrich the uranium in the fuel.

Hazards

Plutonium-239 emits alpha particles to become uranium-235. As an alpha emitter, plutonium-239 is not particularly dangerous as an external radiation source, but if it is ingested or breathed in as dust it is very dangerous and carcinogenic. It has been estimated that a pound (454 grams) of plutonium inhaled as plutonium oxide dust could give cancer to two million people. However, ingested plutonium is by far less dangerous as only a tiny fraction is absorbed in gastrointestinal tract. 800 mg would be unlikely to cause a major health risk as far as radiation is concerned. As a heavy metal, plutonium is also toxic.

Weapons grade plutonium (with greater than 90% Pu-239) is used to make nuclear weapons and has many advantages over other fissile material for that purpose. Lower proportions of Pu-239 would make a reliable weapon design difficult or impossible; this is due to the spontaneous fission (and thus neutron production) of the undesirable Pu-240.

Spent Nuclear Fuel

After nuclear fuel goes through a reactor once, it is called spent fuel. This is a bit of a misnomer because there's still a fair amount of energy left in the fuel. Visually, spent fuel looks the same as the fuel that goes into the reactor as it is contained in fuel assemblies containing ceramic fuel pellets. However, the composition of the fuel pellets has changed a little; some of the uranium atoms within the fuel have split into smaller nuclei. In current reactors only a very small fraction of the uranium has been split, which is why there's so much energy left in the fuel and it's not completely 'spent'.

One wagon of a nuclear waste flask train.

Spent fuel are the fuel bundles that can no longer sustain fission in a nuclear reactor. This happens because having fission daughter products make it too hard to keep the nuclear chain reaction going. These products can be a result of fission or radioactive decays that take place inside the fuel bundle. Most of the waste is solid, but some of the atoms in the spent fuel are gaseous. These isotopes require more careful containment so that they do not escape and cause radiation damage. The waste that comes from nuclear power reactors is known as high level waste or HLW, but there is low level waste as well.

The main concern about this spent fuel is that it is radioactive (and stays radioactive for a long time before decaying to stability), however, this radioactivity is never out in the open and exposed to people (or the surrounding environment). This radioactivity also results in the nuclear fuel emitting heat long after removal from the reactor because of the presence of short-lived fission products. This spent fuel is always shielded, generally by placing the fuel underwater until the radiation is minimized to levels where concrete can act as a shield. After this, there are a couple of disposal options for the spend fuel including storage in a deep geologic repository and nuclear fuel recycling.

Composition

Spent nuclear fuel varies in its composition. This composition somewhat depends on how long the fuel was in the reactor—but a great deal relies on how long the waste has been out of the reactor. When uranium-235 (the fissile isotope in nuclear fuel) is "burned" inside of a reactor, the percentage of the uranium-235 decreases. The other isotope of uranium, uranium-238 doesn't directly undergo much fission but needs an extra neutron before it can split with thermal neutrons (this is referred to as a fertile nucleus). Uranium-238 does become plutonium, neptunium, americium, and curium in these spent fuel bundles.

Spent fuel still has uranium-235, although the amount has fallen. For a light water reactor (like a PWR or BWR) the uranium-235 levels have fallen to about 1%. In CANDU reactors it's even lower. About 95% of the spent fuel is still uranium-238, which can be used again. The other compounds that now exist in the fuel are heavy isotopes (plutonium, americium, neptunium and curium) and other products of fission. Because uranium exists in this waste, it can be used in fast reactors as a fuel. Some of the plutonium produced can even be used as fuel.

Although there are thousands of tons of spent fuel, it is compact and uniform. This means that regardless of what reactor the spent fuel came from it is similar in its composition— making it relatively easy to manage this waste.

Storage and Recycling

Spent fuel can be stored underwater in spent fuel pools near power plants or be buried deep underground. These pools act both as a shield and a coolant for the fuel. Once cooled significantly, the rods are placed in a dry cask storage. These casks allow for air-cooling of the rods, shielding, and are very sturdy. When stored underground, surveying must be done to ensure that the area is geologically stable so that the waste is not disturbed.

Spent fuel contains large amounts of uranium, therefore, lots of usable fuel still remains. This makes recycling this fuel possible, and the process of performing separations on the spent fuel to access this uranium and plutonium is known as reprocessing. It is important to note that reprocessing is much more expensive and difficult than mining uranium and enriching it. To recycle the waste, the spent fuel is chemically processed and used in fast reactors. This results in less nuclear waste and more energy extracted from raw uranium. Currently, France and Japan recycle fuel one time before it is disposed, and the US previously had a recycling program that was shut down because it created plutonium. Plutonium is arguably the easiest material to make a nuclear weapon with, and thus its production is heavily monitored.

Developing molten salt reactors could allow for recycling and breeding without the production of plutonium or other actinides. Fission products are still created, but the reduction of these actinides is one major benefit of using molten salt reactors.

Nuclear Power as Renewable Energy

Whether nuclear power should be considered as a form of renewable energy is an ongoing subject of debate. Statutory definitions of renewable energy usually exclude many present nuclear energy technologies, with notable exceptions in the states of Utah. Dictionary sourced definitions of renewable energy technologies often omit or explicitly exclude mention to every nuclear energy source, with an exception made for the natural nuclear decay heat generated within the Earth/geothermal energy.

The most common fuel used in conventional nuclear fission power stations, uranium-235 is "non-renewable" according to the Energy Information Administration, the organization however is silent on the recycled MOX fuel. Similarly, the National Renewable Energy Laboratory does not mention nuclear power in its "energy basics" definition.

In 1987, the Brundtland Commission (WCED) classified fission reactors that produce more fissile nuclear fuel than they consume (breeder reactors, and if developed, fusion power) among conventional renewable energy sources, such as solar power and hydropower. The American Petroleum Institute likewise does not consider conventional nuclear fission as renewable, but that breeder reactor nuclear fuel is considered renewable and sustainable, and while conventional fission leads to waste streams that remain a concern for millennia, the waste from efficiently burnt up spent fuel requires a more limited storage supervision period of about thousand years. The monitoring and storage of radioactive waste products is also required upon the use of other renewable energy sources, such as geothermal energy.

Renewable Energy

Renewable energy flows involve natural phenomena, which with the exception of tidal power, ultimately derive their energy from the sun (a natural fusion reactor) or from geothermal energy, which is heat derived in greatest part from that which is generated in the earth from the decay of radioactive isotopes, as the International Energy Agency explains:

> "Renewable energy is derived from natural processes that are replenished constantly. In its various forms, it derives directly from the sun, or from heat generated deep within the earth. Included in the definition is electricity and heat generated from sunlight, wind, oceans, hydropower, biomass, geothermal resources, and biofuels and hydrogen derived from renewable resources".

Renewable energy resources exist over wide geographical areas, in contrast to other energy sources, which are concentrated in a limited number of countries.

In ISO 13602-1:2002, a renewable resource is defined as "a natural resource for which the ratio of the creation of the natural resource to the output of that resource from nature to the technosphere is equal to or greater than one".

Conventional Fission and Breeder Reactors as Renewable

Nuclear fission reactors are a natural energy phenomenon, having naturally formed on earth in times past, for example a natural nuclear fission reactor which ran for thousands of years in present-day Oklo Gabon was discovered in the 1970s. It ran for a few hundred thousand years, averaging 100 kW of thermal power during that time.

Conventional, human manufactured, nuclear fission power stations largely use uranium, a common metal found in seawater, and in rocks all over the world, as its primary source of fuel. Uranium-235 "burnt" in conventional reactors, without fuel recycling, is a non-renewable resource, and if used at present rates would eventually be exhausted.

In figure shows a cutaway model of the 2nd most powerful presently operating fast breeder reactor in the world. The (BN-600), at 600 MW of nameplate capacity is equivalent in power output to a natural gas CCGT. It dispatches 560 MW to the Middle Urals power grid. Construction of a second breeder reactor, the BN-800 reactor was completed in 2014.

This is also somewhat similar to the situation with a commonly classified renewable source, geothermal energy, a form of energy derived from the natural nuclear decay of the large, but nonetheless finite supply of uranium, thorium and potassium-40 present within the Earth's crust, and due to the nuclear decay process, this renewable energy source will also eventually run out of fuel. As too will the Sun, and be exhausted.

Nuclear fission involving breeder reactors, a reactor which *breeds* more fissile fuel than they consume and thereby has a breeding ratio for fissile fuel higher than 1 thus has a stronger case for being considered a renewable resource than conventional fission reactors. Breeder reactors would constantly replenish the available supply of nuclear fuel by converting fertile materials, such as uranium-238 and thorium, into fissile isotopes of plutonium or uranium-233, respectively. Fertile materials are also nonrenewable, but their supply on Earth is extremely large, with a supply timeline greater than geothermal energy. In a closed nuclear fuel cycle utilizing breeder reactors, nuclear fuel could therefore be considered renewable.

In 1983, physicist Bernard Cohen claimed that fast breeder reactors, fueled exclusively by natural uranium extracted from seawater, could supply energy at least as long as the sun's expected remaining lifespan of five billion years. This was based on calculations involving the geological cycles of erosion, subduction, and uplift, leading to humans consuming half of the total uranium in the Earth's crust at an annual usage rate of 6500 tonne/yr, which was enough to produce approximately 10 times the world's 1983 electricity consumption, and would reduce the concentration of uranium in the seas by 25%, resulting in an increase in the price of uranium of less than 25%.

Advancements at Oak Ridge National Laboratory and the University of Alabama, as published in a 2012 issue of the American Chemical Society, towards the extraction of uranium from seawater have focused on increasing the biodegradability of the process and reducing the projected cost of the metal if it was extracted from the sea on an industrial scale. The researchers' improvements include using electrospun Shrimp shell Chitin mats that are more effective at absorbing uranium when compared to the prior record setting Japanese method of using plastic amidoxime nets. As of 2013 only a few kilograms (picture available) of uranium have been extracted from the ocean in pilot programs and it is also believed that the uranium extracted on an industrial scale from the seawater would constantly be replenished from uranium leached from the ocean floor, maintaining the seawater concentration at a stable level.

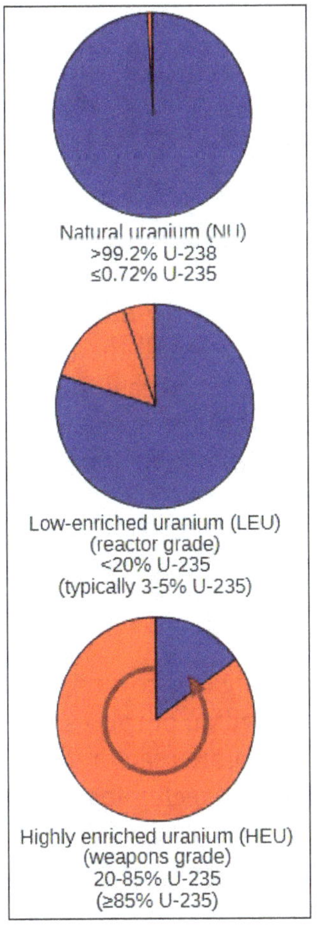

In figure above shows proportions of the isotopes, U-238 (blue) and U-235 (red) found in natural uranium, versus grades that are enriched. light water reactors and the natural uranium capable CANDU reactors, are primarily powered only by the U-235 component, failing to extract much energy from U-238. While by contrast uranium breeder reactors mostly use U-238/the primary constituent of natural uranium as their fuel.

In 1987, the World Commission on Environment and Development(WCED), an organization independent from, but created by, the United Nations, published Our Common Future, in which a particular subset of presently operating nuclear fission technologies, and nuclear fusion were both classified as renewable. That is, fission reactors that produce more fissile fuel than they consume - breeder reactors, and when it is developed, fusion power, are both classified within the same category as conventional renewable energy sources, such as solar and falling water.

Presently, as of 2014, only 2 breeder reactors are producing industrial quantities of electricity, the BN-600 and BN-800. The retired French Phénix reactor also demonstrated a greater than one breeding ratio and operated for ~30 years, producing power when *Our Common Future* was published in 1987. While human sustained nuclear fusion is intended to be proven in the International thermonuclear experimental reactor between 2020 and 2030, and there are also efforts to create a pulsed fusion power reactor based on the inertial confinement principle.

Fusion Fuel Supply

If it is developed, Fusion power would provide more energy for a given weight of fuel than any fuel-consuming energy source currently in use, and the fuel itself (primarily deuterium) exists abundantly in the Earth's ocean: about 1 in 6500 hydrogen (H) atoms in seawater (H_2O) is deuterium in the form of (semi-heavy water). Although this may seem a low proportion (about 0.015%), because nuclear fusion reactions are so much more energetic than chemical combustion and seawater is easier to access and more plentiful than fossil fuels, fusion could potentially supply the world's energy needs for millions of years.

In the deuterium + lithium fusion fuel cycle, 60 million years is the estimated supply lifespan of this fusion power, if it is possible to extract all the lithium from seawater, assuming current world energy consumption. While in the second easiest fusion power fuel cycle, the deuterium + deuterium burn, assuming all of the deuterium in seawater was extracted and used, there is an estimated 150 billion years of fuel, with this again, assuming current world energy consumption.

Nuclear Power Debate

The nuclear power debate is a long-running controversy about the risks and benefits of using nuclear reactors to generate electricity for civilian purposes. The debate about nuclear power peaked during the 1970s and 1980s, as more and more reactors were built and came online, and "reached an intensity unprecedented in the history of technology controversies" in some countries. Thereafter, the nuclear industry created jobs, focused on safety, and public concerns mostly waned. In the last decade, however, with growing public awareness about climate change and the critical role that carbon dioxide and methane emissions plays in causing the heating of the earth's atmosphere, there has been a resurgence in the intensity of the nuclear power debate. Nuclear power advocates and those most concerned about climate change point to nuclear power's reliable, emission-free, high-density energy, alongside a generation of young physicists and engineers working to bring a new generation of nuclear technology into existence to replace fossil fuels. On the other hand, skeptics point to nuclear accidents such as the death of Louis Slotin, the Windscale fire, the Three Mile Island accident, the Chernobyl disaster, and the Fukushima Daiichi nuclear disaster, combined with escalating acts of global terrorism, to argue against continuing use of the technology.

The debate continues today between those who fear the power of nuclear and those who fear what will happen to the earth if humanity doesn't use nuclear power. At the 1963 ground-breaking for what would become the world's largest nuclear power plant, President John F. Kennedy declared that nuclear power was a "step on the long road to peace," and that by using "science and technology to achieve significant breakthroughs" that we could "conserve the resources" to leave the world in better shape. Yet he also acknowledged that the Atomic Age was a "dreadful age" and "when we broke the atom apart, we changed the history of the world."

Proponents of nuclear energy argue that nuclear power is a clean and sustainable energy source which provides huge amounts of uninterrupted energy without polluting the atmosphere or emitting the carbon emissions that cause global warming. Use of nuclear power provides plentiful, well-paying jobs, energy security, reduces a dependence on imported fuels and exposure to price risks associated with resource speculation and Middle East politics. Proponents advance the notion that nuclear power produces virtually no air pollution, in contrast to the massive amount of pollution and carbon emission generated from burning fossil fuels like coal, oil and natural gas. Modern society demands always-on energy to power communications, computer networks, transportation, industry and residences at all times of day and night. In the absence of nuclear power, utilities need to burn fossil fuels to keep the energy grid reliable, even with access to solar and wind energy, because those sources are intermittent. Proponents also believe that nuclear power is the only viable course for a country to achieve energy independence while also meeting their "ambitious" nationally determined contributions (NDCs) to reduce carbon emissions in accordance with the Paris Agreement signed by 195 nations. They emphasize that the risks of storing waste are small and existing stockpiles can be reduced by using this waste to produce fuels for the latest technology in newer reactors. The operational safety record of nuclear is excellent when compared to the other major kinds of power plants and by preventing pollution, actually saves lives every year.

Opponents say that nuclear power poses numerous threats to people and the environment and point to studies in the literature that question if it will ever be a sustainable energy source. These threats include health risks, accidents and environmental damage from uranium mining, processing and transport. Along with the fears associated with nuclear weapons proliferation, nuclear power opponents fear sabotage by terrorists of nuclear plants, diversion and misuse of radioactive fuels or fuel waste, as well as naturally-occurring leakage from the unsolved and imperfect long-term storage process of radioactive nuclear waste. They also contend that reactors themselves are enormously complex machines where many things can and do go wrong, and there have been many serious nuclear accidents. Critics do not believe that these risks can be reduced through new technology. They further argue that when all the energy-intensive stages of the nuclear fuel chain are considered, from uranium mining to nuclear decommissioning, nuclear power is not a low-carbon electricity source.

Electricity and Energy Supplied

The World Nuclear Association has reported that nuclear electricity generation in 2012 was at its lowest level since 1999. The WNA has said that "nuclear power generation suffered its biggest ever one-year fall through 2012 as the bulk of the Japanese fleet remained offline for a full calendar year".

Data from the International Atomic Energy Agency showed that nuclear power plants globally produced 2,346 terawatt-hours (8,450 PJ) of electricity in 2012 – 7% less than in 2011. The figures illustrate the effects of a full year of 48 Japanese power reactors producing no power during the year. The permanent closure of eight reactor units in Germany was also a factor. Problems at Crystal River, Fort Calhoun and the two San Onofre units in the USA meant they produced no power for the full year, while in Belgium Doel 3 and Tihange 2 were out of action for six months. Compared to 2010, the nuclear industry produced 11% less electricity in 2012.

Brazil, China, Germany, India, Japan, Mexico, the Netherlands, Spain and the U.K. now all generate more electricity from non-hydro renewable energy than from nuclear sources. In 2015, new power generation using solar power was 33% of the global total, wind power over 17%, and 1.3% for nuclear power, exclusively due to development in China.

Energy Security

For some countries, nuclear power affords energy independence. Nuclear power has been relatively unaffected by embargoes, and uranium is mined in countries willing to export, including Australia and Canada. However, countries now responsible for more than 30% of the world's uranium production: Kazakhstan, Namibia, Niger, and Uzbekistan, are politically unstable.

One assessment from the IAEA showed that enough high-grade ore exists to supply the needs of the current reactor fleet for 40–50 years. According to Sovacool, reserves from existing uranium mines are being rapidly depleted, and expected shortfalls in available fuel threaten future plants and contribute to volatility of uranium prices at existing plants. Escalation of uranium fuel costs decreased the viability of nuclear projects. Uranium prices rose from 2001 to 2007, before declining.

The International Atomic Energy Agency and the Nuclear Energy Agency of the OCED, in their latest review of world uranium resources and demand, *Uranium 2014: Resources, Production, and Demand*, concluded that uranium resources would support "significant growth in nuclear capacity," and that: "Identified resources are sufficient for over 120 years, considering 2012 uranium requirements of 61 600 tU."

According to a Stanford study, fast breeder reactors have the potential to provide power for humans on earth for billions of years, making this source sustainable. But "because of the link between plutonium and nuclear weapons, the potential application of fast breeders has led to concerns that nuclear power expansion would bring in an era of uncontrolled weapons proliferation".

Reliability

In 2010, the worldwide average capacity factor was 80.1%. In 2005, the global average capacity factor was 86.8%, the number of SCRAMs per 7,000 hours critical was 0.6, and the unplanned capacity loss factor was 1.6%. Capacity factor is the net power produced divided by the maximum amount possible running at 100% all the time, thus this includes all scheduled maintenance/refueling outages as well as unplanned losses. The 7,000 hours is roughly representative of how long any given reactor will remain critical in a year, meaning that the scram rates translates into a sudden and unplanned shut-down about 0.6 times per year for any given reactor in the world. The unplanned capacity loss factor represents amount of power not produced due to unplanned scrams and postponed restarts.

The World Nuclear Association argues that: "Obviously sun, wind, tides and waves cannot be controlled to provide directly either continuous base-load power, or peak-load power when it is needed" "In practical terms non-hydro renewables are therefore able to supply up to some 15–20% of the capacity of an electricity grid, though they cannot directly be applied as economic substitutes for most coal or nuclear power, however significant they become in particular areas with favourable conditions." "If the fundamental opportunity of these renewables is their abundance and relatively widespread occurrence, the fundamental challenge, especially for electricity supply, is applying them to meet demand given their variable and diffuse nature. This means either that there must be reliable duplicate sources of electricity beyond the normal system reserve, or some means of electricity storage." "Relatively few places have scope for pumped storage dams close to where the power is needed, and overall efficiency is less than 80%. Means of storing large amounts of electricity as such in giant batteries or by other means have not been developed."

According to Benjamin K. Sovacool, most studies critiquing solar and wind energy look only at individual generators and not at the system wide effects of solar and wind farms. Correlations between power swings drop substantially as more solar and wind farms are integrated (a process known as geographical smoothing) and a wider geographic area also enables a larger pool of energy efficiency efforts to abate intermittency.

Sovacool says that variable renewable energy sources such as wind power and solar energy can displace nuclear resources. "Nine recent studies have concluded that the variability and intermittency of wind and solar resources becomes easier to manage the more they are deployed and interconnected, not the other way around, as some utilities suggest. This is because wind and solar plants help grid operators handle major outages and contingencies elsewhere in the system, since they generate power in smaller increments that are less damaging than unexpected outages from large plants".

According to a 2011 projection by the International Energy Agency, solar power generators may produce most of the world's electricity within 50 years, with wind power, hydroelectricity and biomass plants supplying much of the remaining generation. "Photovoltaic and concentrated solar power together can become the major source of electricity." Renewable technologies can enhance energy security in electricity generation, heat supply, and transportation.

As of 2013, the World Nuclear Association has said "There is unprecedented interest in renewable energy, particularly solar and wind energy, which provide electricity without giving rise to any carbon dioxide emission. Harnessing these for electricity depends on the cost and efficiency of the technology, which is constantly improving, thus reducing costs per peak kilowatt."

Renewable electricity supply in the 20-50+% range has already been implemented in several European systems, albeit in the context of an integrated European grid system. In 2012 the share of electricity generated by renewable sources in Germany was 21.9%, compared to 16.0% for nuclear power after Germany shut down 7–8 of its 18 nuclear reactors in 2011. In the United Kingdom, the amount of energy produced from renewable energy is expected to exceed that from nuclear power by 2018, and Scotland plans to obtain all electricity from renewable energy by 2020. The majority of installed renewable energy across the world is in the form of hydro power, which has limited opportunity for expansion.

The IPCC has said that if governments were supportive, and the full complement of renewable energy technologies were deployed, renewable energy supply could account for almost 80% of the world's energy use within forty years. Rajendra K. Pachauri, chairman of the IPCC, said the necessary investment in renewables would cost only about 1% of global GDP annually. This approach could contain greenhouse gas levels to less than 450 parts per million, the safe level beyond which climate change becomes catastrophic and irreversible.

The cost of nuclear power has followed an increasing trend whereas the cost of electricity is declining in wind power. As of 2014, the wind industry in the USA is able to produce more power at lower cost by using taller wind turbines with longer blades, capturing the faster winds at higher elevations. This has opened up new opportunities and in Indiana, Michigan, and Ohio, the price of power from wind turbines built 300 feet to 400 feet above the ground can now compete with conventional fossil fuels like coal. Prices have fallen to about 4 cents per kilowatt-hour in some cases and utilities have been increasing the amount of wind energy in their portfolio, saying it is their cheapest option.

From a safety stand point, nuclear power, in terms of lives lost per unit of electricity delivered, is comparable to and in some cases, lower than many renewable energy sources. There is no radioactive spent fuel that needs to be stored or reprocessed with conventional renewable energy sources. A nuclear plant needs to be disassembled and removed. Much of the disassembled nuclear plant needs to be stored as low level nuclear waste.

Since nuclear power plants are fundamentally heat engines, waste heat disposal becomes an issue at high ambient temperature. Droughts and extended periods of high temperature can "cripple nuclear power generation, and it is often during these times when electricity demand is highest because of air-conditioning and refrigeration loads and diminished hydroelectric capacity". In such very hot weather a power reactor may have to operate at a reduced power level or even shut down. In 2009 in Germany, eight nuclear reactors had to be shut down simultaneously on hot summer days for reasons relating to the overheating of equipment or of rivers. Overheated discharge water has resulted in significant killing of fish in the past, harming livelihood and raising public concern.

Economics

New Nuclear Plants

In figure, EDF has said its third-generation EPR Flamanville 3 project will be delayed until 2018, due to "both structural and economic reasons," and the project's total cost has climbed to EUR 11 billion in 2012. Similarly, the cost of the EPR being built at Olkiluoto, Finland has escalated dramatically, and the project is well behind schedule. The initial low cost forecasts for these megaprojects exhibited "optimism bias".

The economics of new nuclear power plants is a controversial subject, since there are diverging views on this topic, and multibillion-dollar investments ride on the choice of an energy source. Nuclear power plants typically have high capital costs for building the plant, but low direct fuel costs (with much of the costs of fuel extraction, processing, use and long term storage externalized). Therefore, comparison with other power generation methods is strongly dependent on assumptions about construction timescales and capital financing for nuclear plants. Cost estimates also need to take into account plant decommissioning and nuclear waste storage costs. On the other hand, measures to mitigate global warming, such as a carbon tax or carbon emissions trading, may favor the economics of nuclear power.

In recent years there has been a slowdown of electricity demand growth and financing has become more difficult, which impairs large projects such as nuclear reactors, with very large upfront costs and long project cycles which carry a large variety of risks. In Eastern Europe, a number of long-established projects are struggling to find finance, notably Belene in Bulgaria and the additional reactors at Cernavoda in Romania, and some potential backers have pulled out. The reliable availability of cheap gas poses a major economic disincentive for nuclear projects.

Analysis of the economics of nuclear power must take into account who bears the risks of future uncertainties. To date all operating nuclear power plants were developed by state-owned or regulated utility monopolies where many of the risks associated with construction costs, operating performance, fuel price, and other factors were borne by consumers rather than suppliers. Many countries have now liberalized the electricity market where these risks, and the risk of cheaper competitors emerging before capital costs are recovered, are borne by plant suppliers and operators rather than consumers, which leads to a significantly different evaluation of the economics of new nuclear power plants.

Following the 2011 Fukushima Daiichi nuclear disaster, costs are likely to go up for currently operating and new nuclear power plants, due to increased requirements for on-site spent fuel management and elevated design basis threats.

Cost of Decommissioning Nuclear Plants

The price of energy inputs and the environmental costs of every nuclear power plant continue long after the facility has finished generating its last useful electricity. Both nuclear reactors and uranium enrichment facilities must be decommissioned, returning the facility and its parts to a safe enough level to be entrusted for other uses. After a cooling-off period that may last as long as a century, reactors must be dismantled and cut into small pieces to be packed in containers for final disposal. The process is very expensive, time-consuming, dangerous for workers, hazardous to the natural environment, and presents new opportunities for human error, accidents or sabotage.

The total energy required for decommissioning can be as much as 50% more than the energy needed for the original construction. In most cases, the decommissioning process costs between US$300 million to US$5.6 billion. Decommissioning at nuclear sites which have experienced a serious accident are the most expensive and time-consuming. In the U.S. there are 13 reactors that have permanently shut down and are in some phase of decommissioning, and none of them have completed the process.

Current UK plants are expected to exceed £73bn in decommissioning costs.

Subsidies

Nuclear power proponents argue that competing energy sources also receive subsidies. Fossil fuels receive large direct and indirect subsidies, such as tax benefits and not having to pay for the greenhouse gases they emit, such as through a carbon tax. Renewable energy sources receive proportionately large direct production subsidies and tax breaks in many nations, although in absolute terms they are often less than subsidies received by non-renewable energy sources.

George W. Bush signing the Energy Policy Act of 2005, which was designed to promote US nuclear reactor construction, through incentives and subsidies, including cost-overrun support up to a total of $2 billion for six new nuclear plants.

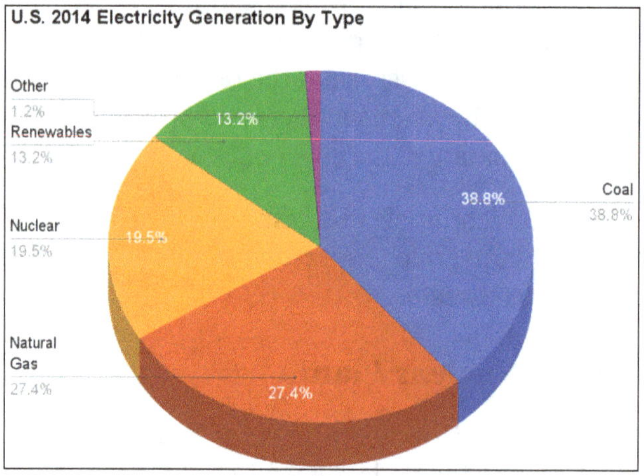

U.S. 2014 Electricity Generation By Type.

Critics of nuclear power claim that it is the beneficiary of inappropriately large economic subsidies, taking the form of research and development, financing support for building new reactors and decommissioning old reactors and waste, and that these subsidies are often overlooked when comparing the economics of nuclear against other forms of power generation.

In Europe, the FP7 research program has more subsidies for nuclear power than for renewable and energy efficiency together; over 70% of this is directed at the ITER fusion project. In the US, public research money for nuclear fission declined from 2,179 to 35 million dollars between 1980 and 2000.

A 2010 report by Global Subsidies Initiative compared relative subsidies of most common energy sources. It found that nuclear energy receives 1.7 US cents per kilowatt hour (kWh) of energy it produces, compared to fossil fuels receiving 0.8 US cents per kWh, renewable energy receiving 5.0 US cents per kWh and biofuels receiving 5.1 US cents per kWh.

Indirect Nuclear Insurance Subsidy

Kristin Shrader-Frechette has said "if reactors were safe, nuclear industries would not demand government-guaranteed, accident-liability protection, as a condition for their generating electricity". No private insurance company or even consortium of insurance companies "would shoulder the fearsome liabilities arising from severe nuclear accidents".

The potential costs resulting from a nuclear accident (including one caused by a terrorist attack or a natural disaster) are great. The liability of owners of nuclear power plants in the U.S. is currently limited under the Price-Anderson Act (PAA). The Price-Anderson Act, introduced in 1957, was "an implicit admission that nuclear power provided risks that producers were unwilling to assume without federal backing". The Price-Anderson Act "shields nuclear utilities, vendors and suppliers against liability claims in the event of a catastrophic accident by imposing an upper limit on private sector liability". Without such protection, private companies were unwilling to be involved. No other technology in the history of American industry has enjoyed such continuing blanket protection.

The PAA was due to expire in 2002, and the former U.S. vice-president Dick Cheney said in 2001 that "nobody's going to invest in nuclear power plants" if the PAA is not renewed.

In 1983, U.S. Nuclear Regulatory Commission (USNRC) concluded that the liability limits placed on nuclear insurance were significant enough to constitute a subsidy, but did not attempt to quantify the value of such a subsidy at that time. Shortly after this in 1990, Dubin and Rothwell were the first to estimate the value to the U.S. nuclear industry of the limitation on liability for nuclear power plants under the Price Anderson Act. Their underlying method was to extrapolate the premiums operators currently pay versus the full liability they would have to pay for full insurance in the absence of the PAA limits. The size of the estimated subsidy per reactor per year was $60 million prior to the 1982 amendments, and up to $22 million following the 1988 amendments. In a separate article in 2003, Anthony Heyes updates the 1988 estimate of $22 million per year to $33 million (2001 dollars).

In case of a nuclear accident, should claims exceed this primary liability, the PAA requires all licensees to additionally provide a maximum of $95.8 million into the accident pool – totaling roughly $10 billion if all reactors were required to pay the maximum. This is still not sufficient in the case of a serious accident, as the cost of damages could exceed $10 billion. According to the PAA, should the costs of accident damages exceed the $10 billion pool, the process for covering the remainder of the costs would be defined by Congress. In 1982, a Sandia National Laboratories study concluded that depending on the reactor size and 'unfavorable conditions' a serious nuclear accident could lead to property damages as high as $314 billion while fatalities could reach 50,000.

Environmental Effects

The primary environmental effects of nuclear power come from uranium mining, radioactive effluent emissions, and waste heat. Nuclear generation does not directly produce sulfur dioxide, nitrogen oxides, mercury or other pollutants associated with the combustion of fossil fuels.

Nuclear plants require slightly more cooling water than fossil-fuel power plants due to their slightly lower generation efficiencies. Uranium mining can use large amounts of water — for example, the Roxby Downs mine in South Australia uses 35 million litres (9,200,000 US gal) of water each day and plans to increase this to 150 million litres (40,000,000 US gal) per day.

Effect on Greenhouse Gas Emissions

While nuclear power does not directly emit greenhouse gases, emissions occur, as with every source of energy, over a facility's life cycle: mining and fabrication of construction materials, plant construction, operation, uranium mining and milling, and plant decommissioning. A literature survey by the Intergovernmental Panel on Climate Change of 32 greenhouse gas emissions studies, found a median value of 16 g (0.56 oz) equivalent lifecycle carbon dioxide emissions per kilowatt hour (kWh) for nuclear power.

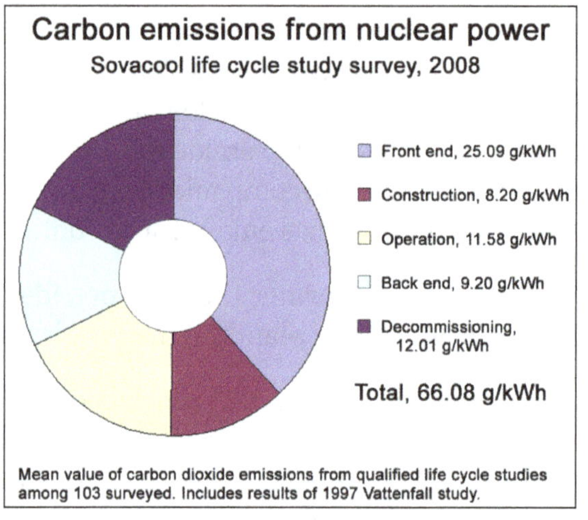

In figure, according to Sovacool, nuclear power plants produce electricity with about 66 g (2.3 oz) equivalent lifecycle carbon dioxide emissions per kWh, while renewable power generators produce electricity with 9.5–38 g (0.34–1.34 oz) carbon dioxide per kWh. A 2012 study estimate, and found that the mean value from nuclear power ranged from 11–25 g/kWh (0.11–0.24 oz/MJ) of total life cycle CO_2 emissions.

Climate and energy scientists James Hansen, Ken Caldeira, Kerry Emanuel and Tom Wigley have released an open letter stating, in part, that:

> "Renewables like wind and solar and biomass will certainly play roles in a future energy economy, but those energy sources cannot scale up fast enough to deliver cheap and reliable power at the scale the global economy requires. While it may be theoretically possible to stabilize the climate without nuclear power, in the real world there is no credible path to climate stabilization that does not include a substantial role for nuclear power".

In a published rebuttal to Hansen's analyses, eight energy and climate scholars say that "nuclear power reactors are less effective at displacing greenhouse gas emissions than energy efficiency initiatives and renewable energy technologies". They go on to argue "that (a) its near-term potential is significantly limited compared to energy efficiency and renewable energy; (b) it displaces emissions and saves lives only at high cost and at the enhanced risk of nuclear weapons proliferation;

(c) it is unsuitable for expanding access to modern energy services in developing countries; and (d) Hansen's estimates of cancer risks from exposure to radiation are flawed". James Hansen and a colleague subsequently wrote a counter-rebuttal.

Mark Diesendorf and B.K. Sovacool review the "little-known research which shows that the life-cycle CO_2 emissions of nuclear power may become comparable with those of fossil power as the 5.4 million tonnes of high-grade uranium ore is used up over the next several decades and low-grade uranium is mined and milled using fossil fuels."

As the nuclear power debate continues, greenhouse gas emissions are increasing. Predictions estimate that even with draconian emission reductions within the ten years, the world will still pass 650ppm of carbon dioxide and a catastrophic 4 °C (7.2 °F) average rise in temperature. Public perception is that renewable energies such as wind, solar, biomass and geothermal are significantly affecting global warming. All of these sources combined only supplied 1.3% of global energy in 2013 as 8 billion tonnes (1.8×10^{13} lb) of coal was burned annually. This "too little, too late" effort may be a mass form of climate change denial, or an idealistic pursuit of green energy.

Another argument is based on a rebound effect—specific to nuclear power energy—on growth and greenhouse gas: it's not the direct effect that would matter but the effect on consumption due to changes in prices and incomes. More research in this field is needed.

High-level Radioactive Waste

Spent nuclear fuel stored underwater and uncapped at the Hanford site in Washington, USA.

The world's nuclear fleet creates about 10,000 metric tons (22,000,000 pounds) of high-level spent nuclear fuel each year. High-level radioactive waste management concerns management and disposal of highly radioactive materials created during production of nuclear power. The technical issues in accomplishing this are daunting, due to the extremely long periods radioactive wastes remain deadly to living organisms. Of particular concern are two long-lived fission products, technetium-99 (half-life 220,000 years) and iodine-129 (half-life 15.7 million years), which dominate spent nuclear fuel radioactivity after a few thousand years. The most troublesome transuranic elements in spent fuel are neptunium-237 (half-life two million years) and plutonium-239 (half-life 24,000 years). Consequently, high-level radioactive waste requires sophisticated treatment and management to successfully isolate it from the biosphere. This usually necessitates treatment, followed by a long-term management strategy involving permanent storage, disposal or transformation of the waste into a non-toxic form. However, many nuclear power by-products are usable as nuclear fuel themselves; extracting the usable energy producing contents from nuclear waste is called "nuclear recycling".

Governments around the world are considering a range of waste management and disposal options, usually involving deep-geologic placement, although there has been limited progress toward implementing long-term waste management solutions. This is partly because the timeframes in question when dealing with radioactive waste range from 10,000 to millions of years, according to studies based on the effect of estimated radiation doses.

Since the fraction of a radioisotope's atoms decaying per unit of time is inversely proportional to its half-life, the relative radioactivity of a quantity of buried human radioactive waste would diminish over time compared to natural radioisotopes (such as the decay chain of 120 trillion tons of thorium and 40 trillion tons of uranium which are at relatively trace concentrations of parts per million each over the crust's 3×10^{19} ton mass).

For instance, over a timeframe of thousands of years, after the most active short half-life radioisotopes decayed, burying U.S. nuclear waste would increase the radioactivity in the top 2,000 feet (610 m) of rock and soil in the United States (100 million km^2 or 39 million sq mi) by approximately 0.1 parts per million over the cumulative amount of natural radioisotopes in such a volume, although the vicinity of the site would have a far higher concentration of artificial radioisotopes underground than such an average.

Nuclear waste disposal is one of the most controversial facets of the nuclear power debate. Presently, waste is mainly stored at individual reactor sites and there are over 430 locations around the world where radioactive material continues to accumulate. Experts agree that centralized underground repositories which are well-managed, guarded, and monitored, would be a vast improvement. There is an international consensus on the advisability of storing nuclear waste in deep underground repositories, but no country in the world has yet opened such a site. There are dedicated waste storage sites at the Waste Isolation Pilot Plant in New Mexico and two in German salt mines, the Morsleben Repository and the Schacht Asse II.

Prevented Mortality

In March 2013, climate scientists Pushker Kharecha and James Hansen published a paper in Environmental Science & Technology, entitled *Prevented mortality and greenhouse gas emissions from historical and projected nuclear power*. It estimated an average of 1.8 million lives saved worldwide by the use of nuclear power instead of fossil fuels between 1971 and 2009. The paper examined mortality levels per unit of electrical energy produced from fossil fuels (coal and natural gas) as well as nuclear power. Kharecha and Hansen assert that their results are probably conservative, as they analyze only deaths and do not include a range of serious but non-fatal respiratory illnesses, cancers, hereditary effects and heart problems, nor do they include the fact that fossil fuel combustion in developing countries tends to have a higher carbon and air pollution footprint than in developed countries. The authors also conclude that the emission of some 64 billion tonnes (7.1×10^{10} tons) of carbon dioxide equivalent have been avoided by nuclear power between 1971 and 2009, and that between 2010 and 2050, nuclear power could additionally avoid up to 80–240 billion tonnes (8.8×10^{10}–2.65×10^{11} tons).

Nuclear Waste

The term nuclear waste is often used to refer to spent nuclear fuel. This term is general, as there are many different types of nuclear waste, but refers to any radioactive waste substance that is produced from industrial processes, including nuclear power plants. Nuclear waste is also produced at hospitals and other medical facilities.

This spent nuclear fuel is placed in a cold pool to cool.

Spent nuclear fuel is specifically high level waste, but it is the waste that many think of. There are many different grades of nuclear waste. Generally, the more radioactive the waste, the less there is of it. The categories of waste are:

- Exempt waste: This type of waste is also referred to as very low level waste. This type of waste is any material that has so little radioactivity, and thus poses no threat. Waste of this kind includes concrete from decommissioned nuclear power plant buildings.
- Low level waste: This type of waste is generated by facilities that use radioactive isotopes. This includes the nuclear power industry, but also includes medical facilities (such as hospitals or radiology clinics) that use medical isotopes to assess and treat diseases like cancer. The radioactivity from low level waste is far less intense than from spent fuel, but the volume is much greater.
- Intermediate level waste: This waste has sufficient radiation that it requires shielding and needs to be handled more carefully than low level waste. There is much less intermediate level waste than low level waste. This type of waste includes the metal around the fuel pellets if the fuel is reprocessed.
- High level waste: This waste is the used fuel and the material that has been drawn out of the used nuclear fuel.

Exempt waste, as the name implies, isn't monitored as radioactive waste, but the rest of the types are. Different levels of waste have different laws proscribing how each level is to be handled.

References

- Taubman, William (2017). Gorbachev: His Life and Times. New York City: Simon and Schuster. P. 291. ISBN 978-1471147968
- Nuclear-physics, entry: newworldencyclopedia.org, Retrieved 27 January, 2019
- "Physical, Nuclear, and Chemical Properties of Plutonium". Institute for Energy and Environmental Research. Retrieved 20 November 2015
- Reactor-physics, nuclear-power: nuclear-power.net, Retrieved 28 February, 2019
- Walker, J. Samuel (10 January 2006). Three Mile Island: A Nuclear Crisis in Historical Perspective. University of California Press. Pp. 10–11. ISBN 9780520246836
- Nuclear-fuel, encyclopedia: energyeducation.ca, Retrieved 29 March, 2019
- "Renewable energy: Definitions from Dictionary.com". Dictionary.com website. Lexico Publishing Group, LLC. Retrieved 2007-08-25
- Spent-nuclear-fuel, Spent-nuclear-fue: energyeducation.ca, Retrieved 30 April, 2019
- Cullati, Stéphane; Kliegel, Matthias; Widmer, Eric (2018). "Development of reserves over the life course and onset of vulnerability in later life". Nature Human Behaviour. 2 (8): 551–558. doi:10.1038/s41562-018-0395-3. ISSN 2397-3374
- Nuclear-waste, encyclopedia: energyeducation.ca, Retrieved 1 May, 2019

3

Nuclear Reactions

Nuclear reaction refers to the change in the identity and traits of an atomic nucleus that are induced by bombarding it with an energetic particle such as alpha particle, neutron, proton, heavy ion, etc. Few significant areas of nuclear reactions are neutron activation, neutron capture, nuclear chain reaction, nuclear fission and nuclear fusion. All these key concepts related to nuclear reactions have been carefully analyzed in this chapter.

A nuclear reaction is considered to be the process in which two nuclear particles (two nuclei or a nucleus and a nucleon) interact to produce two or more nuclear particles or γ-rays (gamma rays). Thus, a nuclear reaction must cause a transformation of at least one nuclide to another. Sometimes if a nucleus interacts with another nucleus or particle without changing the nature of any nuclide, the process is referred to a nuclear scattering, rather than a nuclear reaction. Perhaps the most notable nuclear reactions are the nuclear fusion reactions of light elements that power the energy production of stars and the Sun. Natural nuclear reactions occur also in the interaction between cosmic rays and matter.

The most notable man-controlled nuclear reaction is the fission reaction which occurs in nuclear reactors. Nuclear reactors are devices to initiate and control a nuclear chain reaction, but there are not only manmade devices. The world's first nuclear reactor operated about two billion years ago. The natural nuclear reactor formed at Oklo in Gabon, Africa, when a uranium-rich mineral deposit became flooded with groundwater that acted as a neutron moderator, and a nuclear chain reaction started. These fission reactions were sustained for hundreds of thousands of years, until a chain reaction could no longer be supported. This was confirmed by existence of isotopes of the fission-product gas xenon and by different ratio of U-235/U-238 (enrichment of natural uranium).

Notation of Nuclear Reactions

Standard nuclear notation shows the chemical symbol, the mass number and the atomic number of the isotope.

If the initial nuclei are denoted by *a* and *b*, and the product nuclei are denoted by *c* and *d*, the reaction can be represented by the equation:

$$a + b \rightarrow c + d$$

$${}^{10}_{5}B + {}^{1}_{0}n \rightarrow {}^{7}_{3}Li + {}^{4}_{2}He + 2.8Mev$$

Nuclear Force

Instead of using the full equations in the style above, in many situations a compact notation is used to describe nuclear reactions. This style of the form a(b,c)d is equivalent to a + b producing c + d. Light particles are often abbreviated in this shorthand, typically p means proton, n means neutron, d means deuteron, α means an alpha particle or helium-4, β means beta particle or electron, γ means gamma photon, etc. The reaction above would be written as 10B(n,α)7Li.

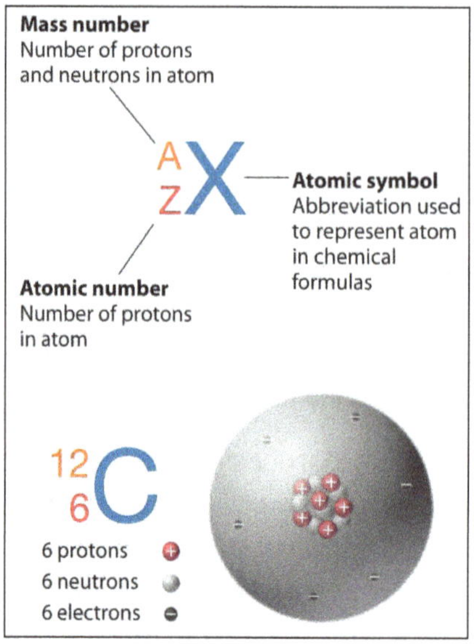

Notation of nuclei.

Basic Classification of Nuclear Reactions

In order to understand the nature of neutron nuclear reactions, the classification according to the time scale of of these reactions has to be introduced. Interaction time is critical for defining the reaction mechanism.

There are two extreme scenarios for nuclear reactions (not only neutron reactions):

- A projectile and a target nucleus are within the range of nuclear forces for the very short time allowing for an interaction of a single nucleon only. These type of reactions are called the direct reactions.
- A projectile and a target nucleus are within the range of nuclear forces for the time allowing for a large number of interactions between nucleons. These type of reactions are called the compound nucleus reactions.

In fact, there is always some non-direct (multiple internuclear interaction) component in all reactions, but the direct reactions have this component limited.

Basic Characteristics of Direct Reactions

- The direct reactions are fast and involve a single-nucleon interaction.
- The interaction time must be very short ($\sim 10^{-22}$ s).

- The direct reactions require incident particle energy larger than ~ 5 MeV/Ap. (Ap is the atomic mass number of a projectile).
- Incident particles interact on the surface of a target nucleus rather than in the volume of a target nucleus.
- Products of the direct reactions are not distributed isotropically in angle, but they are forward focused.
- Direct reactions are of importance in measurements of nuclear structure.

Basic Characteristics of Compound Nucleus Reactions

- The compound nucleus is a relatively long-lived intermediate state of particle-target composite system.
- The compound nucleus reactions involve many nucleon-nucleon interactions.
- The large number of collisions between the nucleons leads to a thermal equilibrium inside the compound nucleus.
- The time scale of compound nucleus reactions is of the order of 10^{-18} s – 10^{-16} s.
- The compound nucleus reactions is usually created if the projectile has low energy.
- Incident particles interact in the volume of a target nucleus.
- Products of the compound nucleus reactions are distributed near isotropically in angle (the nucleus loses memory of how it was created – the Bohr's hypothesis of independence).
- The mode of decay of compound nucleus do not depend on the way the compound nucleus is formed.
- Resonances in the cross-section are typical for the compound nucleus reaction.

Types of Nuclear Reactions

Although the number of possible nuclear reactions is enormous, nuclear reactions can be sorted by types. Most of nuclear reactions are accompanied by gamma emission. Some examples are:

- Elastic scattering: Occurs, when no energy is transferred between the target nucleus and the incident particle.

  208Pb (n, n) 208Pb

- Inelastic scattering: Occurs, when energy is transferred. The difference of kinetic energies is saved in excited nuclide.

  40Ca (α, α') 40mCa

- Capture reactions: Both charged and neutral particles can be captured by nuclei. This is accompanied by the emission of γ-rays. Neutron capture reaction produces radioactive nuclides (induced radioactivity).

238U (n, ˠ) 239U

- Transfer Reactions: The absorption of a particle accompanied by the emission of one or more particles is called the transfer reaction.

  4He (α, p) 7Li

- Fission reactions: Nuclear fission is a nuclear reaction in which the nucleus of an atom splits into smaller parts (lighter nuclei). The fission process often produces free neutrons and photons (in the form of gamma rays), and releases a large amount of energy.

  235U (n, 3 n) fission products

- Fusion reactions: Occur when, two or more atomic nuclei collide at a very high speed and join to form a new type of atomic nucleus.The fusion reaction of deuterium and tritium is particularly interesting because of its potential of providing energy for the future.

  3T (d, n) 4He

- Spallation reactions: Occur, when a nucleus is hit by a particle with sufficient energy and momentum to knock out several small fragments or, smash it into many fragments.

- Nuclear decay (Radioactive decay): Occurs when an unstable atom loses energy by emitting ionizing radiation. Radioactive decay is a random process at the level of single atoms, in that, according to quantum theory, it is impossible to predict when a particular atom will decay. There are many types of radioactive decay-

  - Alpha radioactivity: Alha particles consist of two protons and two neutrons bound together into a particle identical to a helium nucleus. Because of its very large mass (more than 7000 times the mass of the beta particle) and its charge, it heavy ionizes material and has a very short range.

  - Beta radioactivity: Beta particles are high-energy, high-speed electrons or positrons emitted by certain types of radioactive nuclei such as potassium-40. The beta particles have greater range of penetration than alpha particles, but still much less than gamma rays.The beta particles emitted are a form of ionizing radiation also known as beta rays. The production of beta particles is termed beta decay.

  - Gamma radioactivity: Gamma rays are electromagnetic radiation of an very high frequency and are therefore high energy photons. They are produced by the decay of nuclei as they transition from a high energy state to a lower state known as gamma decay. Most of nuclear reactions are accompanied by gamma emission.

- Neutron emission: Neutron emission is a type of radioactive decay of nuclei containing excess neutrons (especially fission products), in which a neutron is simply ejected from the nucleus. This type of radiation plays key role in nuclear reactor control, because these neutrons are delayed neutrons.

Decay Type	Radiation Emitted	Generic Equation	Model
Alpha decay	$^{4}_{2}\alpha$	$^{A}_{Z}X \longrightarrow ^{A-4}_{Z-2}X' + ^{4}_{2}\alpha$	Parent → Daughter, Alpha Particle
Beta decay	$^{0}_{-1}\beta$	$^{A}_{Z}X \longrightarrow ^{A}_{Z+1}X' + ^{0}_{-1}\beta$	Parent → Daughter, Beta Particle
Positron emission	$^{0}_{+1}\beta$	$^{A}_{Z}X \longrightarrow ^{A}_{Z-1}X' + ^{0}_{+1}\beta$	Parent → Daughter, Positron
Electron capture	X rays	$^{A}_{Z}X + ^{0}_{-1}e \longrightarrow ^{A}_{Z-1}X' +$ X ray	Parent, Electron → Daughter, X ray
Gamma emission	$^{0}_{0}\gamma$	$^{A}_{Z}X^{*} \xrightarrow{\text{Relaxation}} ^{A}_{Z}X' + ^{0}_{0}\gamma$	Parent (excited nuclear state) → Daughter, Gamma ray
Spontaneous fission	Neutrons	$^{A+B+C}_{Z+Y}X \longrightarrow ^{A}_{Z}X' + ^{B}_{Y}X' + C^{1}_{0}n$	Parent (unstable) → Daughters, ENERGY, Neutrons

Conservation Laws in Nuclear Reactions

In analyzing nuclear reactions, we apply the many conservation laws. Nuclear reactions are subject to classical conservation laws for charge, momentum, angular momentum, and energy(including rest energies). Additional conservation laws, not anticipated by classical physics, are:

- Law of Conservation of Lepton Number.
- Law of Conservation of Baryon Number.
- Law of Conservation of Electric Charge.

Law of Conservation of Lepton Number

In particle physics, the lepton number is used to denote which particles are leptons and which particles are not. Each lepton has a lepton number of 1 and each antilepton has a lepton number of -1. Other non-leptonic particles have a lepton number of 0. The lepton number is a conserved quantum number in all particle reactions. A slight asymmetry in the laws of physics allowed leptons to be created in the Big Bang.

The conservation of lepton number means that whenever a lepton of a certain generation is created or destroyed in a reaction, a corresponding antilepton from the same generation must be created or destroyed. It must be added, there is a separate requirement for each of the three generations of leptons, the electron, muon and tau and their associated neutrinos.

Q/e	$L_e=-1$	$L_\mu=-1$	$L_\tau=-1$
0 +1	$\begin{pmatrix}\overline{\nu}_e\\ e^+\end{pmatrix}$	$\begin{pmatrix}\overline{\nu}_\mu\\ \mu^+\end{pmatrix}$	$\begin{pmatrix}\overline{\nu}_\tau\\ \tau^+\end{pmatrix}$
Q/e	$L_e=1$	$L_\mu=1$	$L_\tau=1$
0 −1	$\begin{pmatrix}\nu_e\\ e^-\end{pmatrix}$	$\begin{pmatrix}\nu_\mu\\ \mu^-\end{pmatrix}$	$\begin{pmatrix}\nu_\tau\\ \tau^-\end{pmatrix}$

Consider the decay of the neutron. The reaction involves only first generation leptons: electrons and neutrinos:

$$n \rightarrow p + e^- + \overline{\nu_e}$$

$$\textit{Lepton Number}: 0 \rightarrow 0 + 1 - 1$$

Since the lepton number must be equal to zero on both sides and it was found that the reaction is a three-particle decay (the electrons emitted in beta decay have a continuous rather than a discrete spectrum), the third particle must be an electron antineutrino.

Law of Conservation of Baryon Number

In particle physics, the baryon number is used to denote which particles are baryons and which particles are not. Each baryon has a baryon number of 1 and each antibaryon has a baryon number of -1. Other non-baryonic particles have a baryon number of 0. Since there are exotic hadrons like pentaquarks and tetraquarks, there is a general definition of baryon number as:

$$B = 1/3\ (n_q - n_{\bar{q}})$$

where, n_q is the number of quarks, and n_q is the number of antiquarks.

The baryon number is a conserved quantum number in all particle reactions.

The law of conservation of baryon number states that:

The sum of the baryon number of all incoming particles is the same as the sum of the baryon numbers of all particles resulting from the reaction.

For example, the following reaction has never been observed:

$$p + n \rightarrow p + p + \bar{p}$$

$$\textit{Baryon Number}: 1 + 1 \rightarrow 1 + 1 - 1$$

Even if the incoming proton has sufficient energy and charge, energy, and so on, are conserved. This reaction does not conserve baryon number since the left side has B =+2, and the right has B =+1.

On the other hand, the following reaction (proton-antiproton pair production) does conserve B and does occur if the incoming proton has sufficient energy (the threshold energy = 5.6 GeV):

$$p + n \rightarrow p + p + \overline{p} + p$$

$$\textit{Baryon Number}: 1 + 1 \rightarrow 1 + 1 - 1 + 1$$

As indicated, B = +2 on both sides of this equation.

From these and other reactions, the conservation of baryon number has been established as a basic principle of physics.

This principle provides basis for the stability of the proton. Since the proton is the lightest particle among all baryons, the hypothetical products of its decay would have to be non-baryons. Thus, the decay would violate the conservation of baryon number. It must be added some theories have suggested that protons are in fact unstable with very long half-life (~10^{30} years) and that they decay into leptons. There is currently no experimental evidence that proton decay occurs.

Law of Conservation of Electric Charge

The law of conservation of electric charge can be demonstrated also on positron-electron pair production. Since a gamma ray is electrically neutral and sum of the electric charges of electron and positron is also zero, the electric charge in this reaction is also conserved.

$$\gamma \rightarrow e^- + e^+$$

It must be added, in order for electron-positron pair production to occur, the electromagnetic energy of the photon must be above a threshold energy, which is equivalent to the rest mass of two electrons. The threshold energy (the total rest mass of produced particles) for electron-positron pair production is equal to 1.02MeV (2 x 0.511MeV) because the rest mass of a single electron is equivalent to 0.511MeV of energy. If the original photon's energy is greater than 1.02MeV, any energy above 1.02MeV is according to the conservation law split between the kinetic energy of motion of the two particles. The presence of an electric field of a heavy atom such as lead or uranium is essential in order to satisfy conservation of momentum and energy. In order to satisfy both conservation of momentum and energy, the atomic nucleus must receive some momentum. Therefore a photon pair production in free space cannot occur.

Certain of these laws are obeyed under all circumstances, others are not. We have accepted conservation of energy and momentum. In all the examples given we assume that the number of protons and the number of neutrons is separately conserved. We shall find circumstances and conditions in which this rule is not true. Where we are considering non-relativistic nuclear reactions, it is essentially true. However, where we are considering relativistic nuclear energies or those involving the weak interactions, we shall find that these principles must be extended.

Some conservation principles have arisen from theoretical considerations, others are just empirical relationships. Notwithstanding, any reaction not expressly forbidden by the conservation laws will generally occur, if perhaps at a slow rate. This expectation is based on quantum mechanics. Unless the barrier between the initial and final states is infinitely high, there is always a non-zero probability that a system will make the transition between them.

For purposes of analyzing non-relativistic reactions, it is sufficient to note four of the fundamental laws governing these reactions:

1. Conservation of nucleons: The total number of nucleons before and after a reaction are the same.

2. Conservation of charge: The sum of the charges on all the particles before and after a reaction are the same.

3. Conservation of momentum: The total momentum of the interacting particles before and after a reaction are the same.
4. Conservation of energy: Energy, including rest mass energy, is conserved in nuclear reactions.

Example: Elastic Nuclear Collision.

A neutron (n) of mass 1.01 u traveling with a speed of 3.60 x 10^4m/s interacts with a carbon (C) nucleus ($m_C = 12.00$ u) initially at rest in an elastic head-on collision.

What are the velocities of the neutron and carbon nucleus after the collision?

Solution:

This is an elastic head-on collision of two objects with unequal masses. We have to use the conservation laws of momentum and of kinetic energy, and apply them to our system of two particles.

Conservation of Momentum

total P before=total P after

$$m_n v_n + 0 = m_n v_n' m_c v_c'$$

Conservation of Kinetic Energy

total k before = total K after

$$\frac{1}{2} m_n v_n^2 + 0 = \frac{1}{2} m_n v_n'^2 + \frac{1}{2} m_c v_c'^2$$

This equation stated that the relative speed of the two objects after the collision has the same magnitude (but opposite direction) as before the collision, no matter what the masses are.

$$v_A + v_A' = v_B' + v_B$$

In our case:

$$v_n - 0 = v_c' - v_n'$$

thus,

$$v_n' = v_c' - v_n$$

Substituting this to our momentum, equation, we get:

$$m_n v_n = m_n v_c' - m_n v_n + m_c v_c'$$

Solving this equation for both velocities, we obtaion:

$$v_c' = \frac{2m_n v_n}{m_n + m_c} = \frac{2 \ x \ 1.01}{13.01} \ x \ 3.6 \ x \ 10^4 m/s$$

$$v_n' = v_c' - v_n = 0.56 \ x \ 10^4 - 3.6 \ x \ 10^4 = -3.04 \ x \ 10^4 \ m/s$$

The minus sign for v' tells us that the neutron scatters back of the carbon nucleus, because the carbon nucleus is significantly heavier. On the other hand its speed is less than its initial speed. This process is known as the neutron moderation and it significantly depends on the mass of moderator nuclei.

Energetics of Nuclear Reactions - Q-value

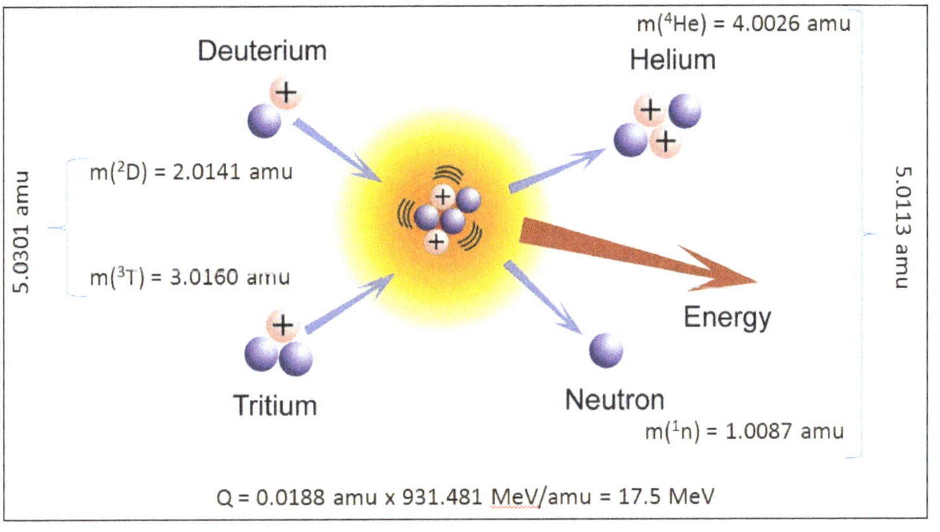

In nuclear and particle physics the energetics of nuclear reactions is determined by the Q-value of that reaction. The Q-value of the reaction is defined as the difference between the sum of the masses of the initial reactants and the sum of the masses of the final products, in energy units (usually in MeV).

Consider a typical reaction, in which the projectile a and the target A gives place to two products, B and b. This can also be expressed in the notation that we used so far, a + A → B + b, or even in a more compact notation, A(a,b)B.

The Q-value of this reaction is given by:

$$Q = [m_a + m_A - (m_b + m_B)]c^2$$

which is the same as the excess kinetic energy of the final products:

$$Q = T_{final} - T_{initial}$$

$$= T_b + T_B - (T_a + T_A)$$

For reactions in which there is an increase in the kinetic energy of the products Q is positive. The positive Q reactions are said to be exothermic (or exergic). There is a net release of energy, since the kinetic energy of the final state is greater than the kinetic energy of the initial state.

For reactions in which there is a decrease in the kinetic energy of the products Q is negative. The negative Q reactions are said to be endothermic (or endoergic) and they require a net energy input.

The energy released in a nuclear reaction can appear mainly in one of three ways:

- Kinetic energy of the products.

- Emission of gamma rays: Gamma rays are emitted by unstable nuclei in their transition from a high energy state to a lower state known as gamma decay.
- Metastable state: Some energy may remain in the nucleus, as a metastable energy level.

A small amount of energy may also emerge in the form of X-rays. Generally, products of nuclear reactions may have different atomic numbers, and thus the configuration of their electron shells is different in comparison with reactants. As the electrons rearrange themselves and drop to lower energy levels, internal transition X-rays (X-rays with precisely defined emission lines) may be emitted.

Exothermic Reactions

Exothermic Reaction - DT fusion

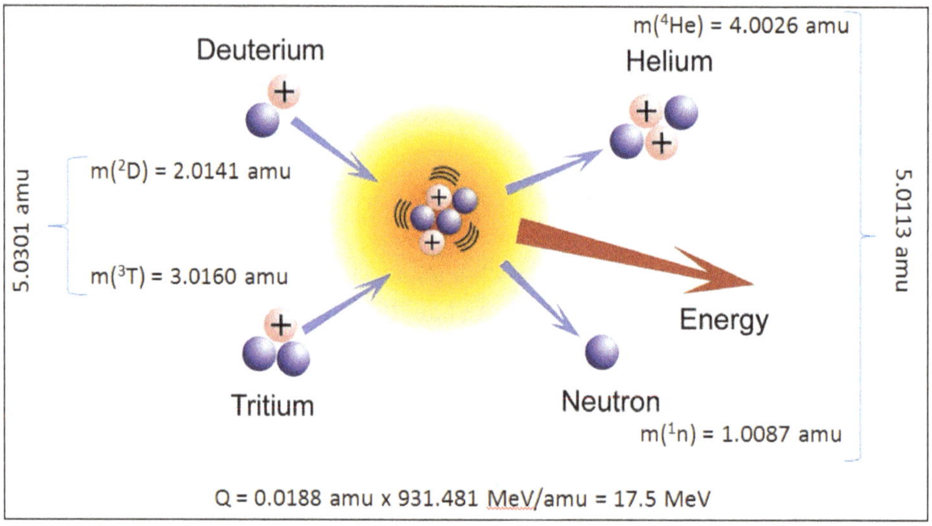

Q-value of DT fusion reaction.

The DT fusion reaction of deuterium and tritium is particularly interesting because of its potential of providing energy for the future. Calculate the reaction Q-value.

3T (d, n) 4He

The atom masses of the reactants and products are:

- $m(^3T) = 3.0160$ amu
- $m(^2D) = 2.0141$ amu
- $m(^1n) = 1.0087$ amu
- $m(^4He) = 4.0026$ amu

Using the mass-energy equivalence, we get the Q-value of this reaction as:

$$Q = \{(3.0160+2.0141)\,[amu] - (1.0087+4.0026)\,[amu]\} \times 931.481\,[MeV/amu]$$

$$= 0.0188 \times 931.481 = 17.5\ MeV$$

Example: Exothermic reaction - tritium in reactors.

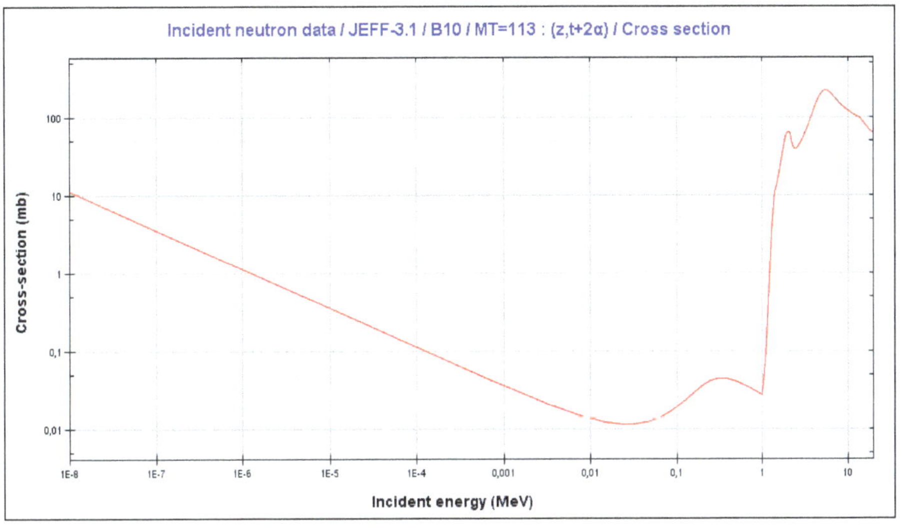

Cross-section of 10B(n,2alpha)T reaction.

Tritium is a byproduct in nuclear reactors. Most of the tritium produced in nuclear power plants stems from the boric acid, which is commonly used as a chemical shim to compensate an excess of initial reactivity. Main reaction, in which the tritium is generated from boron is below:

10B(n,2*alpha)T

This reaction of a neutron with an isotope ^{10}B is the main way, how radioactive tritium in primary circuit of all PWRs is generated. Note that, this reaction is a threshold reaction due to its cross-section.

Calculate the reaction Q-value.

The atom masses of the reactants and products are:

- $m(^{10}B) = 10.01294$ amu
- $m(^{1}n) = 1.00866$ amu
- $m(^{3}T) = 3.01604$ amu
- $m(^{4}He) = 4.0026$ amu

Using the mass-energy equivalence, we get the Q-value of this reaction as:

$$Q = \{(10.0129+1.00866)\ [amu] - (3.01604+2 \times 4.0026)\ [amu]\} \times 931.481\ [MeV/amu]$$

$$= 0.00036 \times 931.481 = 0.335\ MeV$$

Endothermic Reactions

Endothermic Reaction – Photoneutrons

In nuclear reactors the gamma radiation plays a significant role also in reactor kinetics and in a subcriticality control. Especially in nuclear reactors with D_2O moderator (CANDU reactors) or with Be reflectors (some experimental reactors). Neutrons can be produced also in (γ, n) reactions and therefore they are usually referred to as photoneutrons.

A high energy photon (gamma ray) can under certain conditions eject a neutron from a nucleus. It occurs when its energy exceeds the binding energy of the neutron in the nucleus. Most nuclei have binding energies in excess of 6 MeV, which is above the energy of most gamma rays from fission. On the other hand there are few nuclei with sufficiently low binding energy to be of practical interest. These are: ^{2}D, ^{9}Be, ^{6}Li, ^{7}Li and ^{13}C. As can be seen from the table the lowest threshold have ^{9}Be with 1.666 MeV and ^{2}D with 2.226 MeV.

Nuclide	Threshold(MeV)	Reaction
^{2}D	2.225	$^{2}H(\gamma,n)\ ^{1}H$
^{6}Li	3.697	$^{6}Li(\gamma,n+p)\ ^{4}He$
^{6}Li	5.67	$^{6}Li(\gamma,n)\ ^{5}Li$
^{7}Li	7.251	$^{7}Li(\gamma,n)\ ^{5}Li$
^{9}Be	1.667	$^{9}Be(\gamma,n)\ ^{8}Be$
^{13}C	4.9	$^{13}C(\gamma,n)\ ^{12}C$

In case of deuterium, neutrons can be produced by the interaction of gamma rays (with a minimum energy of 2.22 MeV) with deuterium:

$$^{2}_{1}D+\gamma \rightarrow {}^{1}_{1}H+n$$

The reaction Q-value is calculated below-

The atom masses of the reactant and products are:

- m(^{2}D) = 2.01363 amu
- m(^{1}n) = 1.00866 amu
- m(^{1}H) = 1.00728 amu

Using the mass-energy equivalence, we get the Q-value of this reaction as:

Q = {2.01363 [amu] – (1.00866+1.00728) [amu]} x 931.481 [MeV/amu]

= -0.00231 x 931.481 = -2.15 MeV

Endothermic Reaction - (α, n) Reaction

Calculate the reaction Q-value of the following reaction:

7Li (α, n) 10B

The atom masses of the reactants and products are:

- m(^{4}He) = 4.0026 amu
- m(^{7}Li) = 7.0160 amu

- $m(^1n) = 1.0087$ amu
- $m(^{10}B) = 10.01294$ amu

Using the mass-energy equivalence, we get the Q-value of this reaction as:

$$Q = \{(7.0160+4.0026)\ [\text{amu}] - (1.0087+10.01294)\ [\text{amu}]\} \times 931.481\ [\text{MeV/amu}]$$

$$= 0.00304 \times 931.481 = -2.83\ \text{MeV}$$

Nuclear Force

The nuclear force (or nucleon–nucleon interaction or residual strong force) is a force that acts between the protons and neutrons of atoms. Neutrons and protons, both nucleons, are affected by the nuclear force almost identically. Since protons have charge +1 *e*, they experience an electric force that tends to push them apart, but at short range the attractive nuclear force is strong enough to overcome the electromagnetic force. The nuclear force binds nucleons into atomic nuclei.

The nuclear force is powerfully attractive between nucleons at distances of about 1 femtometre (fm, or 1.0×10^{-15} metres), but it rapidly decreases to insignificance at distances beyond about 2.5 fm. At distances less than 0.7 fm, the nuclear force becomes repulsive. This repulsive component is responsible for the physical size of nuclei, since the nucleons can come no closer than the force allows. By comparison, the size of an atom, measured in angstroms (Å, or 1.0×10^{-10} m), is five orders of magnitude larger. The nuclear force is not simple, however, since it depends on the nucleon spins, has a tensor component, and may depend on the relative momentum of the nucleons. The strong nuclear force is one of the fundamental forces of nature.

The nuclear force plays an essential role in storing energy that is used in nuclear power and nuclear weapons. Work (energy) is required to bring charged protons together against their electric repulsion. This energy is stored when the protons and neutrons are bound together by the nuclear force to form a nucleus. The mass of a nucleus is less than the sum total of the individual masses of the protons and neutrons. The difference in masses is known as the mass defect, which can be expressed as an energy equivalent. Energy is released when a heavy nucleus breaks apart into two or more lighter nuclei. This energy is the electromagnetic potential energy that is released when the nuclear force no longer holds the charged nuclear fragments together.

A quantitative description of the nuclear force relies on equations that are partly empirical. These equations model the internucleon potential energies, or potentials. (Generally, forces within a system of particles can be more simply modeled by describing the system's potential energy; the negative gradient of a potential is equal to the vector force.) The constants for the equations are phenomenological, that is, determined by fitting the equations to experimental data. The internucleon potentials attempt to describe the properties of nucleon–nucleon interaction. Once determined, any given potential can be used in, e.g., the Schrödinger equation to determine the quantum mechanical properties of the nucleon system.

The discovery of the neutron in 1932 revealed that atomic nuclei were made of protons and neutrons, held together by an attractive force. By 1935 the nuclear force was conceived to be transmitted by particles called mesons. This theoretical development included a description of the Yukawa potential, an early example of a nuclear potential. Mesons, predicted by theory, were discovered experimentally in 1947. By the 1970s, the quark model had been developed, by which the mesons and nucleons were viewed as composed of quarks and gluons. By this new model, the nuclear force, resulting from the exchange of mesons between neighboring nucleons, is a residual effect of the strong force.

While the nuclear force is usually associated with nucleons, more generally this force is felt between hadrons, or particles composed of quarks. At small separations between nucleons (less than ~ 0.7 fm between their centers, depending upon spin alignment) the force becomes repulsive, which keeps the nucleons at a certain average separation, even if they are of different types. This repulsion arises from the Pauli exclusion force for identical nucleons (such as two neutrons or two protons). A Pauli exclusion force also occurs between quarks of the same type within nucleons, when the nucleons are different (a proton and a neutron, for example).

Field Strength

At distances larger than 0.7 fm the force becomes attractive between spin-aligned nucleons, becoming maximal at a center–center distance of about 0.9 fm. Beyond this distance the force drops exponentially, until beyond about 2.0 fm separation, the force is negligible. Nucleons have a radius of about 0.8 fm.

At short distances (less than 1.7 fm or so), the attractive nuclear force is stronger than the repulsive Coulomb force between protons; it thus overcomes the repulsion of protons within the nucleus. However, the Coulomb force between protons has a much greater range as it varies as the inverse square of the charge separation, and Coulomb repulsion thus becomes the only significant force between protons when their separation exceeds about 2 to 2.5 fm.

The nuclear force has a spin-dependent component. The force is stronger for particles with their spins aligned than for those with their spins anti-aligned. If two particles are the same, such as two neutrons or two protons, the force is not enough to bind the particles, since the spin vectors of two particles of the same type must point in opposite directions when the particles are near each other and are (save for spin) in the same quantum state. This requirement for fermions stems from the Pauli exclusion principle. For fermion particles of different types, such as a proton and neutron, particles may be close to each other and have aligned spins without violating the Pauli exclusion principle, and the nuclear force may bind them (in this case, into a deuteron), since the nuclear force is much stronger for spin-aligned particles. But if the particles' spins are anti-aligned the nuclear force is too weak to bind them, even if they are of different types.

The nuclear force also has a tensor component which depends on the interaction between the nucleon spins and the angular momentum of the nucleons, leading to deformation from a simple spherical shape.

Nuclear Binding

To disassemble a nucleus into unbound protons and neutrons requires work against the nuclear force. Conversely, energy is released when a nucleus is created from free nucleons or other nuclei: the nuclear binding energy. Because of mass–energy equivalence (i.e. Einstein's famous formula $E = mc^2$), releasing this energy causes the mass of the nucleus to be lower than the total mass of the individual nucleons, leading to the so-called "mass defect".

The nuclear force is nearly independent of whether the nucleons are neutrons or protons. This property is called *charge independence*. The force depends on whether the spins of the nucleons are parallel or antiparallel, as it has a non-central or *tensor* component. This part of the force does not conserve orbital angular momentum, which under the action of central forces is conserved.

The symmetry resulting in the strong force, proposed by Werner Heisenberg, is that protons and neutrons are identical in every respect, other than their charge. This is not completely true, because neutrons are a tiny bit heavier, but it is an approximate symmetry. Protons and neutrons are therefore viewed as the same particle, but with different isospin quantum numbers; conventionally, the proton is *isospin up*, while the neutron is *isospin down*. The strong force is invariant under SU(2) isospin transformations, just as other interactions between particles are invariant under SU(2) transformations of intrinsic spin. In other words, both isospin and intrinsic spin transformations are isomorphic to the SU(2) symmetry group. There are only strong attractions when the total isospin of the set of interacting particles is 0, which is confirmed by experiment.

Our understanding of the nuclear force is obtained by scattering experiments and the binding energy of light nuclei.

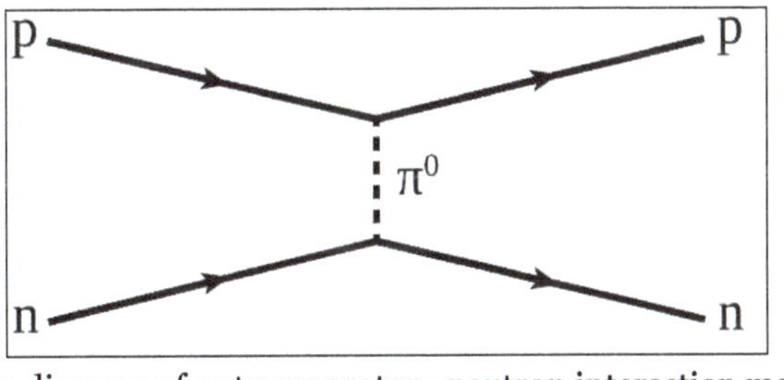

A Feynman diagram of a strong proton–neutron interaction mediated by a neutral pion. Time proceeds from left to right.

The nuclear force occurs by the exchange of virtual light mesons, such as the virtual pions, as well as two types of virtual mesons with spin (vector mesons), the rho mesons and the omega mesons. The vector mesons account for the spin-dependence of the nuclear force in this "virtual meson" picture.

The nuclear force is distinct from what historically was known as the weak nuclear force. The weak interaction is one of the four fundamental interactions, and plays a role in processes such as beta decay. The weak force plays no role in the interaction of nucleons, though it is responsible for the decay of neutrons to protons and vice versa.

Nuclear Force as a Residual of the Strong Force

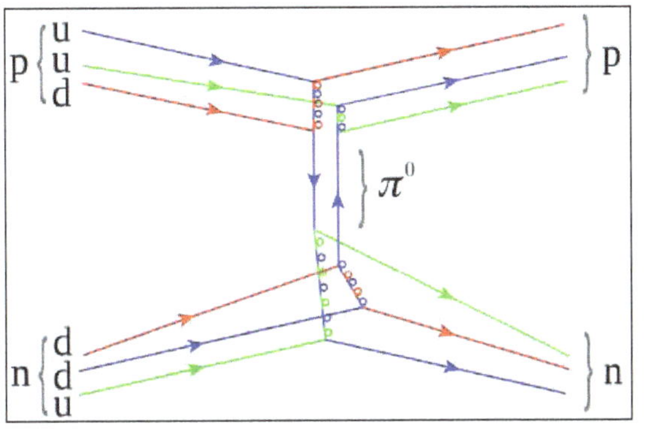

In figure, the same diagram as that above with the individual quark constituents shown, to illustrate how the *fundamental* strong interaction gives rise to the nuclear force. Straight lines are quarks, while multi-colored loops are gluons (the carriers of the fundamental force). Other gluons, which bind together the proton, neutron, and pion "in flight", are not shown.

The nuclear force is a residual effect of the more fundamental strong force, or strong interaction. The strong interaction is the attractive force that binds the elementary particles called quarks together to form the nucleons (protons and neutrons) themselves. This more powerful force is mediated by particles called gluons. Gluons hold quarks together through color charge which is analogous to electric charge, but far stronger. Quarks, gluons, and their dynamics are mostly confined within nucleons, but residual influences extend slightly beyond nucleon boundaries to give rise to the nuclear force.

The nuclear forces arising between nucleons are analogous to the forces in chemistry between neutral atoms or molecules called London forces. Such forces between atoms are much weaker than the attractive electrical forces that hold the atoms themselves together (i.e., that bind electrons to the nucleus), and their range between atoms is shorter, because they arise from small separation of charges inside the neutral atom. Similarly, even though nucleons are made of quarks in combinations which cancel most gluon forces (they are "color neutral"), some combinations of quarks and gluons nevertheless leak away from nucleons, in the form of short-range nuclear force fields that extend from one nucleon to another nearby nucleon. These nuclear forces are very weak compared to direct gluon forces ("color forces" or strong forces) inside nucleons, and the nuclear forces extend only over a few nuclear diameters, falling exponentially with distance. Nevertheless, they are strong enough to bind neutrons and protons over short distances, and overcome the electrical repulsion between protons in the nucleus.

Sometimes, the nuclear force is called the residual strong force, in contrast to the strong interactions which arise from QCD. This phrasing arose during the 1970s when QCD was being established. Before that time, the *strong nuclear force* referred to the inter-nucleon potential. After the verification of the quark model, *strong interaction* has come to mean QCD.

Nucleon–nucleon Potentials

Two-nucleon systems such as the deuteron, the nucleus of a deuterium atom, as well as proton–proton or neutron–proton scattering are ideal for studying the *NN* force. Such systems can be described by attributing a *potential* (such as the Yukawa potential) to the nucleons and using the potentials in a Schrödinger equation. The form of the potential is derived phenomenologically (by measurement), although for the long-range interaction, meson-exchange theories help to construct the potential. The parameters of the potential are determined by fitting to experimental data such as the deuteron binding energy or *NN* elastic scattering cross sections (or, equivalently in this context, so-called *NN* phase shifts).

The most widely used *NN* potentials are the Paris potential, the Argonne AV18 potential, the CD-Bonn potential and the Nijmegen potentials.

A more recent approach is to develop effective field theories for a consistent description of nucleon–nucleon and three-nucleon forces. Quantum hadrodynamics is an effective field theory of the nuclear force, comparable to QCD for color interactions and QED for electromagnetic interactions.

Additionally, chiral symmetry breaking can be analyzed in terms of an effective field theory (called chiral perturbation theory) which allows perturbative calculations of the interactions between nucleons with pions as exchange particles.

From Nucleons to Nuclei

The ultimate goal of nuclear physics would be to describe all nuclear interactions from the basic interactions between nucleons. This is called the *microscopic* or *ab initio* approach of nuclear physics. There are two major obstacles to overcome before this dream can become reality:

- Calculations in many-body systems are difficult and require advanced computation techniques.
- There is evidence that three-nucleon forces (and possibly higher multi-particle interactions) play a significant role. This means that three-nucleon potentials must be included into the model.

This is an active area of research with ongoing advances in computational techniques leading to better first-principles calculations of the nuclear shell structure. Two- and three-nucleon potentials have been implemented for nuclides up to $A = 12$.

Nuclear Potentials

A successful way of describing nuclear interactions is to construct one potential for the whole nucleus instead of considering all its nucleon components. This is called the *macroscopic* approach. For example, scattering of neutrons from nuclei can be described by considering a plane wave in the potential of the nucleus, which comprises a real part and an imaginary part. This model is often called the optical model since it resembles the case of light scattered by an opaque glass sphere.

Nuclear potentials can be *local* or *global*: local potentials are limited to a narrow energy range and a narrow nuclear mass range, while global potentials, which have more parameters and are usually less accurate, are functions of the energy and the nuclear mass and can therefore be used in a wider range of applications.

Nuclear Binding Energy and Mass Defect

- Nuclear binding energy is the energy required to split a nucleus of an atom into its components.
- Nuclear binding energy is used to determine whether fission or fusion will be a favorable process.
- The mass defect of a nucleus represents the mass of the energy binding the nucleus, and is the difference between the mass of a nucleus and the sum of the masses of the nucleons of which it is composed.

Terms:

- Nucleonone of the subatomic particles of the atomic nucleus, i.e. a proton or a neutron.
- Mass defect the difference between the calculated mass of the unbound system and the experimentally measured mass of the nucleus.
- Strong force the nuclear force, a residual force responsible for the interactions between nucleons, deriving from the color force.

Binding Energy

Nuclear binding energy is the energy required to split a nucleus of an atom into its component parts: protons and neutrons, or, collectively, the nucleons. The binding energy of nuclei is always a positive number, since all nuclei require net energy to separate them into individual protons and neutrons.

Mass Defect

Nuclear binding energy accounts for a noticeable difference between the actual mass of an atom's nucleus and its expected mass based on the sum of the masses of its non-bound components.

Recall that energy (E) and mass (m) are related by the equation:

$$E = mc^2$$

Here, c is the speed of light. In the case of nuclei, the binding energy is so great that it accounts for a significant amount of mass.

The actual mass is always less than the sum of the individual masses of the constituent protons and neutrons because energy is removed when when the nucleus is formed. This energy has mass, which is removed from the total mass of the original particles. This mass, known as the mass defect, is missing in the resulting nucleus and represents the energy released when the nucleus is formed.

Mass defect (M_d) can be calculated as the difference between observed atomic mass (m_o) and that expected from the combined masses of its protons (m_p, each proton having a mass of 1.00728 amu) and neutrons (m_n, 1.00867 amu):

$$M_d = (m_n + m_p) - m_o$$

Nuclear Binding Energy

Once mass defect is known, nuclear binding energy can be calculated by converting that mass to energy by using $E=mc^2$. Mass must be in units of kg.

Once this energy, which is a quantity of joules for one nucleus, is known, it can be scaled into per-nucleon and per-mole quantities. To convert to joules/mole, simply multiply by Avogadro's number. To convert to joules per nucleon, simply divide by the number of nucleons.

Nuclear binding energy can also apply to situations when the nucleus splits into fragments composed of more than one nucleon; in these cases, the binding energies for the fragments, as compared to the whole, may be either positive or negative, depending on where the parent nucleus and the daughter fragments fall on the nuclear binding energy curve. If new binding energy is available when light nuclei fuse, or when heavy nuclei split, either of these processes result in the release of the binding energy. This energy—available as nuclear energy—can be used to produce nuclear power or build nuclear weapons. When a large nucleus splits into pieces, excess energy is emitted as photons, or gamma rays, and as kinetic energy, as a number of different particles are ejected.

Nuclear binding energy is also used to determine whether fission or fusion will be a favorable process. For elements lighter than iron-56, fusion will release energy because the nuclear binding energy increases with increasing mass. Elements heavier than iron-56 will generally release energy upon fission, as the lighter elements produced contain greater nuclear binding energy. As such, there is a peak at iron-56 on the nuclear binding energy curve.

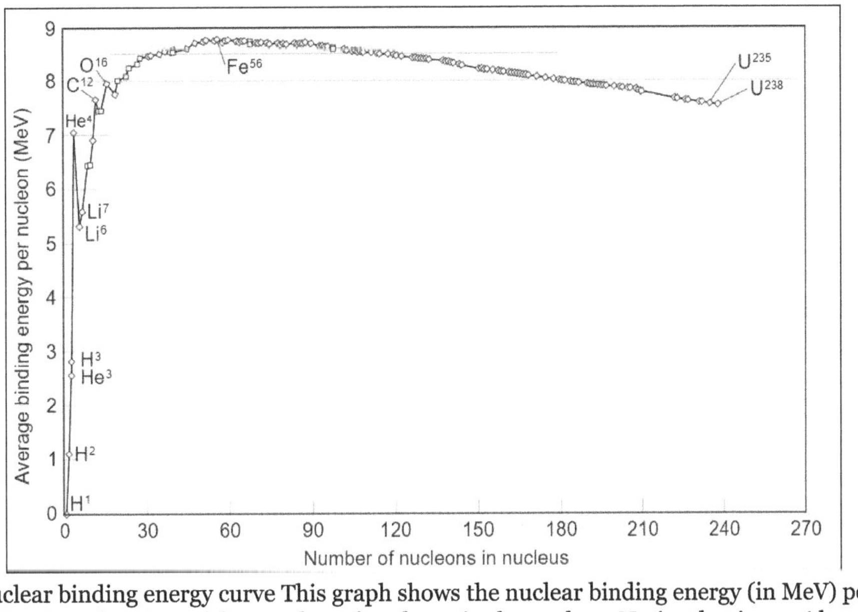

Nuclear binding energy curve This graph shows the nuclear binding energy (in MeV) per nucleon as a function of the number of nucleons in the nucleus. Notice that iron-56 has the most binding energy per nucleon, making it the most stable nucleus.

The rationale for this peak in binding energy is the interplay between the coulombic repulsion of the protons in the nucleus, because like charges repel each other, and the strong nuclear force, or strong force. The strong force is what holds protons and neutrons together at short distances. As the size of the nucleus increases, the strong nuclear force is only felt between nucleons that are close together, while the coulombic repulsion continues to be felt throughout the nucleus; this leads to instability and hence the radioactivity and fissile nature of the heavier elements.

Neutron Activation

Neutron activation is the process in which neutron radiation induces radioactivity in materials, and occurs when atomic nuclei capture free neutrons, becoming heavier and entering excited states. The excited nucleus often decays immediately by emitting gamma rays, or particles such as beta particles, alpha particles, fission products, and neutrons (in nuclear fission). Thus, the process of neutron capture, even after any intermediate decay, often results in the formation of an unstable activation product. Such radioactive nuclei can exhibit half-lives ranging from small fractions of a second to many years.

Neutron activation is the only common way that a stable material can be induced into becoming intrinsically radioactive. All naturally occurring materials, including air, water, and soil, can be induced (activated) by neutron capture into some amount of radioactivity in varying degrees, as a result of production of neutron-rich radioisotopes. Some atoms require more than one neutron to become unstable, which makes them harder to activate because the probability of a double or triple capture by a nucleus is below that of single capture. Water, for example, is made up of hydrogen and oxygen. Hydrogen requires a double capture to attain instability as tritium (hydrogen-3), while natural oxygen (oxygen-16) requires three captures to become unstable oxygen-19. Thus water is relatively difficult to activate, as compared to sea salt (NaCl), in which both the sodium and chlorine ions become unstable with a single capture each. These facts were realized first-hand at the *Operation Crossroads* atomic test series in 1946.

Examples:

An example of this kind of a nuclear reaction occurs in the production of cobalt-60 within a nuclear reactor: The cobalt-60 then decays by the emission of a beta particle plus gamma rays into nickel-60. This reaction has a half-life of about 5.27 years; and due to the availability of cobalt-59 (100% of its natural abundance) this neutron bombarded isotope of cobalt is a valuable source of nuclear radiation (namely gamma radiation) for radiotherapy.

In other cases, and depending on the kinetic energy of the neutron, the capture of a neutron can cause nuclear fission—the splitting of the atomic nucleus into two smaller nuclei. If the fission requires an input of energy, that comes from the kinetic energy of the neutron. An example of this kind of fission in a light element can occur when the stable isotope of lithium, lithium-7, is bombarded with fast neutrons and undergoes the following nuclear reaction:

$$^{7}_{3}\text{Li} + ^{1}_{0}\text{n} \rightarrow ^{4}_{2}\text{He} + ^{3}_{1}\text{H} + ^{1}_{0}\text{n} + \text{gamma rays} + \text{kinetic energy}$$

In other words, the capture of a neutron by lithium-7 causes it to split into an energetic helium nucleus (alpha particle), a hydrogen-3 (tritium) nucleus and a free neutron. The Castle Bravo accident, in which the thermonuclear bomb test at Enewetak Atoll in 1954 exploded with 2.5 times the expected yield, was caused by the unexpectedly high probability of this reaction.

In the areas around a pressurized water reactors or boiling water reactors during normal operation, a significant amount of radiation is produced due to the fast neutron activation of coolant water oxygen via a (n,p) reaction. The activated oxygen-16 nucleus emits a proton (hydrogen nucleus), and transmutes to nitrogen-16, which has a very short life (7.13 seconds) before decaying back to oxygen-16 (emitting 6.13MeV beta particles).

$$^{16}_{8}\text{O} + ^{1}_{0}\text{n} \rightarrow ^{1}_{1}\text{p} + ^{16}_{7}\text{N}(\text{Decays rapidly})$$

$$^{59}_{27}\text{Co} + ^{1}_{0}\text{n} \rightarrow ^{60}_{27}\text{Co}$$

$$^{16}_{7}\text{N} \rightarrow \gamma + ^{0}_{-1}\text{e}^{-} + ^{16}_{8}\text{O}$$

This activation of the coolant water requires extra biological shielding around the nuclear reactor plant. It is the high energy gamma ray in the second reaction that causes the major concern. This is why water that has recently been inside a nuclear reactor core must be shielded until this radiation subsides. One to two minutes is generally sufficient.

In facilities that housed a cyclotron, the concrete foundation can become radioactive due to neutron activation. Six important long-lived radioactive isotopes (^{54}Mn, ^{55}Fe, ^{60}Co, ^{65}Zn, ^{133}Ba, and ^{152}Eu) can be found within concrete nuclei affected by neutrons. The residual radioactivity is predominantly due to trace elements presence and thus the amount of radioactivity derived from cyclotron activation is minuscule, i.e., pCi/g or Bq/g. The release limit for facilities with residual radioactivity is 25 mrem/year. An example of ^{55}Fe production from iron rebar activation is shown below:

$$^{54}_{26}Fe + ^{1}_{0}n \rightarrow ^{55}_{26}Fe$$

Occurrence

Neutron activation is the only common way that a stable material can be induced into becoming intrinsically radioactive. Neutrons are only free in quantity in the microseconds of a nuclear weapon's explosion, in an active nuclear reactor, or in a spallation neutron source.

In an atomic weapon neutrons are only generated for from 1 to 50 microseconds, but in huge numbers. Most are absorbed by the metallic bomb casing, which is only just starting to be affected by the explosion within it. The neutron activation of the soon-to-be vaporized metal is responsible for a significant portion of the nuclear fallout in nuclear bursts high in the atmosphere. In other types of activation neutrons may irradiate soil that is dispersed in a mushroom cloud at or near the Earth's surface, resulting in fallout from activation of soil chemical elements.

Effects on Materials Over Time

In any location with high neutron fluxes, such as within the cores of nuclear reactors, neutron activation contributes to material erosion; periodically the lining materials themselves must be disposed of, as low-level radioactive waste. Some materials are more subject to neutron activation than others, so a suitably chosen low-activation material can significantly reduce this problem . For example, Chromium-51 will form by neutron activation in chrome steel (which contains Cr-50) that is exposed to a typical reactor neutron flux.

Carbon-14, most frequently but not solely, generated by the neutron activation of atmospheric nitrogen-14 with thermal neutron, is (together with its dominant natural production pathway from cosmic ray-air interactions and historical production from atmospheric nuclear testing) also generated in comparatively minute amounts inside many designs of nuclear reactors which contain nitrogen gas impurities in their fuel cladding, coolant water and by neutron activation of the oxygen contained in the water itself. Fast breeder reactors (FBR) produce about an order of magnitude less C-14 than the most common reactor type, the pressurized water reactor, as FBRs do not use water as a primary coolant.

Uses

Radiation Safety

For physicians and radiation safety officers, activation of sodium in the human body to sodium-24, and phosphorus to phosphorus-32, can give a good immediate estimate of acute accidental neutron exposure.

Neutron Detection

One way to demonstrate that nuclear fusion has occurred inside a fusor device is to use a Geiger counter to measure the gamma ray radioactivity that is produced from a sheet of aluminium foil.

In the ICF fusion approach, the fusion yield of the experiment (directly proportional to neutron production) is usually determined by measuring the gamma-ray emissions of aluminium or copper neutron activation targets. Aluminium can capture a neutron and generate radioactive sodium-24, which has a half life of 15 hours and a beta decay energy of 5.514 MeV.

The activation of a number of test target elements such as sulfur, copper, tantalum, and gold have been used to determine the yield of both pure fission and thermonuclear weapons.

Materials Analysis

Neutron activation analysis is one of the most sensitive and accurate methods of trace element analysis. It requires no sample preparation or solubilization and can therefore be applied to objects that need to be kept intact such as a valuable piece of art. Although the activation induces radioactivity in the object, its level is typically low and its lifetime may be short, so that its effects soon disappear. In this sense, neutron activation is a non-destructive analysis method.

Neutron activation analysis can be done in situ. For example, aluminium (Al-27) can be activated by capturing relatively low-energy neutrons to produce the isotope Al-28, which decays with a half-life of 2.3 minutes with a decay energy of 4.642 MeV. This activated isotope is used in oil drilling to determine the clay content (clay is generally an alumino-silicate) of the underground area under exploration.

Historians can use accidental neutron activation to authenticate atomic artifacts and materials subjected to neutron fluxes from fission incidents. For example, one of the fairly unique isotopes found in trinitite, and therefore with its absence likely signifying a fake sample of the mineral, is a barium neutron activation product, the barium in the Trinity device coming from the slow explosive lens employed in the device, known as Baratol.

Neutron Capture

Neutron capture is a nuclear reaction in which an atomic nucleus and one or more neutrons collide and merge to form a heavier nucleus. Since neutrons have no electric charge, they can enter a nucleus more easily than positively charged protons, which are repelled electrostatically.

Neutron capture plays an important role in the cosmic nucleosynthesis of heavy elements. In stars it can proceed in two ways: as a rapid (r-process) or a slow process (s-process). Nuclei of masses greater than 56 cannot be formed by thermonuclear reactions (i.e. by nuclear fusion), but can be formed by neutron capture. Neutron capture on protons yields a line at 2.223 MeV predicted and commonly observed in solar flares.

Neutron Capture at Small Neutron Flux

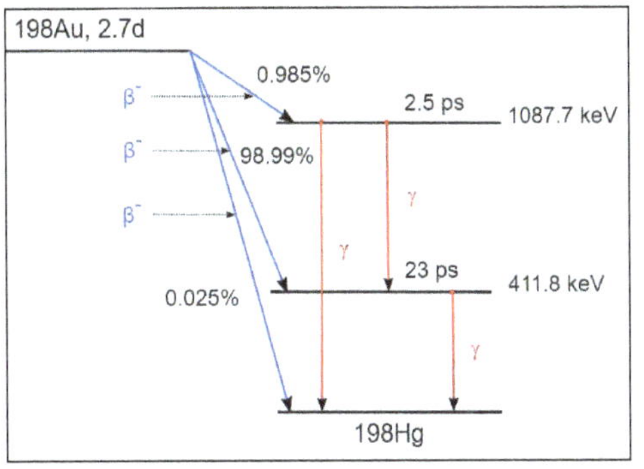

Decay scheme of ^{198}Au.

At small neutron flux, as in a nuclear reactor, a single neutron is captured by a nucleus. For example, when natural gold (^{197}Au) is irradiated by neutrons (n), the isotope ^{198}Au is formed in a highly excited state, and quickly decays to the ground state of ^{198}Au by the emission of gamma rays (γ). In this process, the mass number increases by one. This is written as a formula in the form $^{197}\text{Au}+\text{n} \rightarrow {}^{198}\text{Au}+\gamma$, or in short form $^{197}\text{Au}(\text{n},\gamma)^{198}\text{Au}$. If thermal neutrons are used, the process is called thermal capture.

The isotope ^{198}Au is a beta emitter that decays into the mercury isotope ^{198}Hg. In this process the atomic number rises by one.

Neutron Capture at High Neutron Flux

The r-process happens inside stars if the neutron flux density is so high that the atomic nucleus has no time to decay via beta emission in between neutron captures. The mass number therefore rises by a large amount while the atomic number (i.e., the element) stays the same. When further neutron capture is no longer possible, the highly unstable nuclei decay via many β^- decays to beta-stable isotopes of higher-numbered elements.

Capture Cross-section

The absorption neutron cross-section of an isotope of a chemical element is the effective cross sectional area that an atom of that isotope presents to absorption, and is a measure of the probability of neutron capture. It is usually measured in barns (b).

Absorption cross section is often highly dependent on neutron energy. As a generality, the likelihood of absorption is proportional to the time the neutron is in the vicinity of the nucleus. The time spent in the vicinity of the nucleus is inversely proportional to the relative velocity between the neutron and nucleus. Other more specific issues modify this general principle. Two of the most commonly specified measures are the cross-section for thermal neutron absorption, and resonance integral which considers the contribution of absorption peaks at certain neutron energies specific to a particular nuclide, usually above the thermal range, but encountered as neutron moderation slows the neutron down from an original high energy.

The thermal energy of the nucleus also has an effect; as temperatures rise, Doppler broadening increases the chance of catching a resonance peak. In particular, the increase in uranium-238's ability to absorb neutrons at higher temperatures (and to do so without fissioning) is a negative feedback mechanism that helps keep nuclear reactors under control.

Thermochemical Significance

Neutron capture is involved in the formation of isotopes of chemical elements. As a consequence of this fact the energy of neutron capture intervenes in the standard enthalpy of formation of isotopes.

Uses

Neutron activation analysis can be used to remotely detect the chemical composition of materials. This is because different elements release different characteristic radiation when they absorb neutrons. This makes it useful in many fields related to mineral exploration and security.

Neutron Absorbers

The most important neutron absorber is ^{10}B as $^{10}B_4C$ in control rods, or boric acid as a coolant water additive in PWRs. Other important neutron absorbers that are used in nuclear reactors are xenon, cadmium, hafnium, gadolinium, cobalt, samarium, titanium, dysprosium, erbium, europium, molybdenum and ytterbium; all of which usually consist of mixtures of various isotopes—some of which are excellent neutron-absorbers. These also occur in combinations such as Mo_2B_5, hafnium diboride, titanium diboride, dysprosium titanate and gadolinium titanate.

Hafnium, one of the last stable elements to be discovered, presents an interesting case. Even though hafnium is a heavier element, its electron configuration makes it practically identical with the element zirconium, and they are always found in the same ores. However, their nuclear properties are different in a profound way. Hafnium absorbs neutrons avidly (Hf absorbs 600 times more than Zr), and it can be used in reactor control rods, whereas natural zirconium is practically transparent to neutrons. So, zirconium is a very desirable construction material for reactor internal parts, including the metallic cladding of the fuel rods which contain either uranium, plutonium, or mixed oxides of the two elements (MOX fuel).

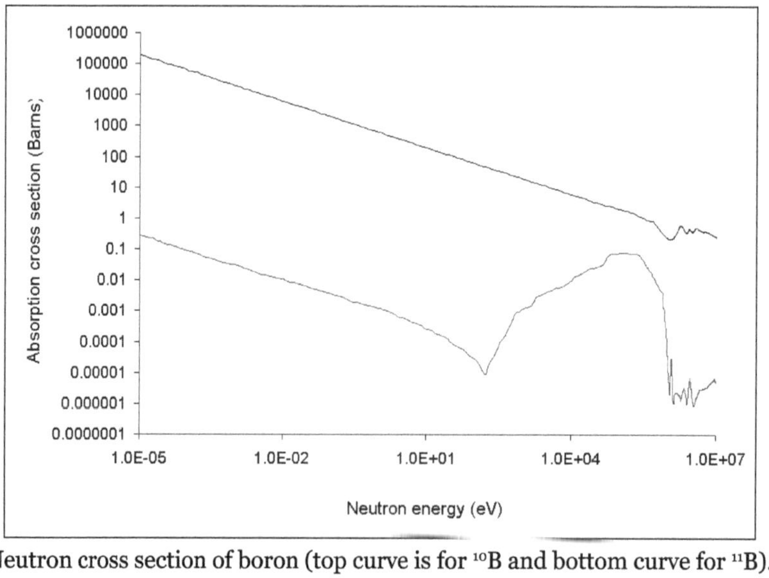

Neutron cross section of boron (top curve is for ^{10}B and bottom curve for ^{11}B).

Hence, it is quite important to be able to separate the zirconium from the hafnium in their naturally occurring alloy. This can only be done inexpensively by using modern chemical ion-exchange resins. Similar resins are also used in reprocessing nuclear fuel rods, when it is necessary to separate uranium and plutonium, and sometimes thorium.

Neutron Cross Section

In nuclear and particle physics, the concept of a neutron cross section is used to express the likelihood of interaction between an incident neutron and a target nucleus. In conjunction with the neutron flux, it enables the calculation of the reaction rate, for example to derive the thermal power of a nuclear power plant. The standard unit for measuring the cross section is the barn, which is equal to 10^{-28} m² or 10^{-24} cm². The larger the neutron cross section, the more likely a neutron will react with the nucleus.

An isotope (or nuclide) can be classified according to its neutron cross section and how it reacts to an incident neutron. Nuclides that tend to absorb a neutron and either decay or keep the neutron in its nucleus are neutron absorbers and will have a *capture cross section* for that reaction. Isotopes that fission, are fissionable fuels and have a corresponding *fission cross section*. The remaining isotopes will simply scatter the neutron, and have a *scatter cross section*. Some isotopes, like uranium-238, have nonzero cross sections of all three.

Isotopes which have a large scatter cross section and a low mass are good neutron moderators . Nuclides which have a large absorption cross section are neutron poisons if they are neither fissile nor undergo decay. A poison that is purposely inserted into a nuclear reactor for controlling its reactivity in the long term and improve its shutdown margin is called a *burnable* poison.

Parameters of Interest

The neutron cross section, and therefore the probability of an interaction, depends on:

- The target type (hydrogen, uranium, etc).
- The type of nuclear reaction (scattering, fission, etc).
- The incident particle energy, also called speed or temperature (thermal, fast, etc).
- To a lesser extent, of:
    - Its relative angle between the incident neutron and the target nuclide.
    - The target nuclide temperature.

Target Type Dependence

The neutron cross section is defined for a given type of target particle. For example, the capture cross section of hydrogen-2 is much smaller than that of common hydrogen-1. This is the reason why some reactors use heavy water (in which most of the hydrogen is deuterium) instead of ordinary light water as moderator: fewer neutrons are lost by capture inside the medium, hence enabling the use of natural uranium instead of enriched uranium. This is the principle of a CANDU reactor.

Type of Reaction Dependence

The likelihood of interaction between an incident neutron and a target nuclide, independent of the type of reaction, is expressed with the help of the total cross section σ_T. However, it may be useful to know if the incoming particle bounces off the target (and therefore continue travelling after the interaction) or disappears after the reaction. For that reason, the scattering and absorption cross sections σ_S and σ_A are defined and the total cross section is simply the sum of the two partial cross sections:

$$\sigma_T = \sigma_S + \sigma_A$$

Absorption Cross-section

If the neutron is absorbed when approaching the nuclide, the atomic nucleus moves up on the table of isotopes by one position. For instance, U-235 becomes U-236* with the * indicating the nucleus is highly energized. This energy has to be released and the release can take place through any of several mechanisms.

- The simplest way for the release to occur is for the neutron to be ejected by the nucleus. If the neutron is emitted immediately, it acts the same as in other scattering events.
- The nucleus may emit gamma radiation.
- The nucleus may β^- decay, where a neutron is converted into a proton, an electron and an electron-type antineutrino (the antiparticle of the neutrino).
- About 81% of the U-236* nuclei are so energized that they undergo fission, releasing the energy as kinetic motion of the fission fragments, also emitting between one and five free neutrons.
- Nuclei that undergo fission as their predominant decay method after neutron capture include U-233, U-235, U-237, Pu-239, Pu-241.
- Nuclei that predominantly absorb neutrons and then emit Beta particle radiation lead to these isotopes, e.g., Th-232 absorbs a neutron and becomes Th-233*, which emits a Beta particle and becomes Pa-233, which emits another Beta particle to become U-233.
- Isotopes that undergo Beta emission transmute from one element to another element, those that undergo gamma or X-ray emission don't change in element or isotope.

Scattering Cross-section

The scattering cross-section can be further subdivided into coherent scattering and incoherent scattering, which is caused by the spin dependence of the scattering cross-section and, for a natural sample, presence of different isotopes of the same element in the sample.

Because neutrons interact with the nuclear potential, the scattering cross-section varies for different isotopes of the element in question. A very prominent example is hydrogen and its isotope deuterium. The total cross-section for hydrogen is over 10 times that of deuterium, mostly due to the large incoherent scattering length of hydrogen. Metals tend to be rather transparent to neutrons, aluminum and zirconium being the two best examples of this.

Incident Particle Energy Dependence

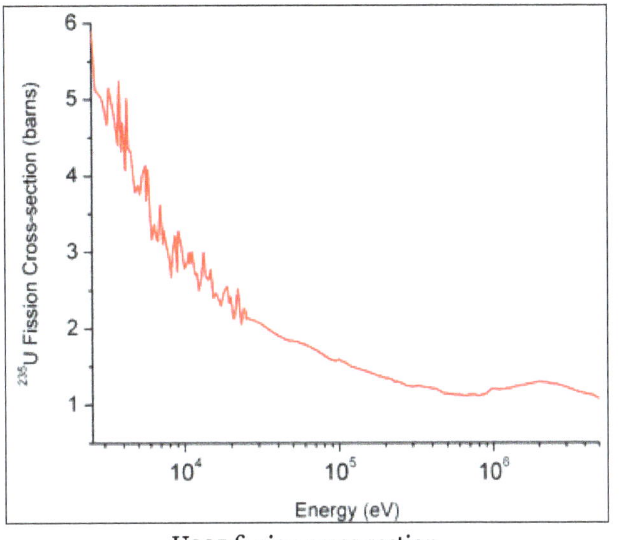

U235 fission cross section.

For a given target and reaction, the cross section is strongly dependent on the neutron speed. In the extreme case, the cross section can be, at low energies, either zero (the energy for which the cross section becomes significant is called threshold energy) or much larger than at high energies.

Therefore, a cross section should be defined either at a given energy or should be averaged in an energy range (or group).

As an example, the plot on the right shows that the fission cross section of the uranium 235 is low at high neutron energies but becomes higher at low energies. Such physical constraint explains why most operational nuclear reactors use a neutron moderator to reduce the energy of the neutron and thus increase the probability of fission, essential to produce energy and sustain the chain reaction.

A simple estimation of energy dependence of any kind of cross section is provided by the Ramsauer Model, which is based on idea that the *effective* size of a neutron is proportional to the breadth of the probability density function of where the neutron is likely to be, which itself is proportional to the neutron's thermal de Broglie wavelength.

$$\lambda(E) = \frac{h}{\sqrt{2mE}}$$

Taking λ as effective radius of the neutron, we can estimate area of circle σ in which neutron hit nuclei of effective radius R as:

$$\sigma(E) \propto \pi(R + \lambda(E))^2$$

While the assumptions of this model are naive, it explains at least qualitatively typical measured energy dependence of neutron absorption cross section. For neutron of wavelength much larger than typical radius of atomic nuclei (1–10 fm, E = 10–1000 keV) the R can be neglected. For these low energy neutrons (such as thermal neutrons) cross section $\sigma(E)$ is inversely proportional to neutron velocity.

This explains the advantage of using neutron moderator in fission nuclear reactor. On the other hand, for very high energy neutrons (over 1 MeV), λ can be neglected, and neutron cross section is approximately constant, determined just by cross section of atomic nuclei.

However, this simple model does not take into account so called neutron resonances, which strongly modify neutron cross section in energy range of 1 eV–10 keV, nor threshold energy of some nuclear reactions.

Target Temperature Dependence

Cross sections are usually measured at 20 °C. To account for the dependence with temperature of the medium (viz. the target), the following formula is used:

$$\sigma = \sigma_0 \left(\frac{T_0}{T} \right)^{\frac{1}{2}},$$

Where, σ is the cross section at temperature T, and σ_0 the cross section at temperature T_0 (T and T_0 in kelvins).

The energy is defined at the most likely energy and velocity of the neutron. The neutron population consists of a Maxwellian distribution, and hence the mean energy and velocity will be higher.

Doppler Broadening

A Doppler broadening of neutron resonances is very important phenomenon, which improves nuclear reactor stability. The prompt temperature coefficient of most thermal reactors is negative, owing to a nuclear Doppler effect. Nuclei are located in atoms which are themselves in continual motion owing to their thermal energy (temperature). As a result of these thermal motions, neutrons impinging on a target appears to the nuclei in the target to have a continuous spread in energy. This, in turn, has an effect on the observed shape of resonance. The resonance becomes shorter and wider than when the nuclei are at rest.

Although the shape of resonances changes with temperature, the total area under the resonance remains essentially constant. But this does not imply constant neutron absorption. Despite the constant area under resonance a resonance integral, which determines the absorption, increases with increasing target temperature. This, of course, decreases coefficient k (negative reactivity is inserted).

Link to Reaction Rate and Interpretation

Imagine a spherical target (outlined in grey in the figure) and a beam of particles (in blue) "flying" at speed v (vector in blue) in the direction of the target. We want to know how many particles impact it during time interval dt. To achieve it, the particles have to be in the green cylinder in the figure (volume V). The base of the cylinder is the geometrical cross section of the target perpendicular to the beam (surface σ in red) and its height the length travelled by the particles during dt (length $v\,dt$):

$$V = \sigma v dt$$

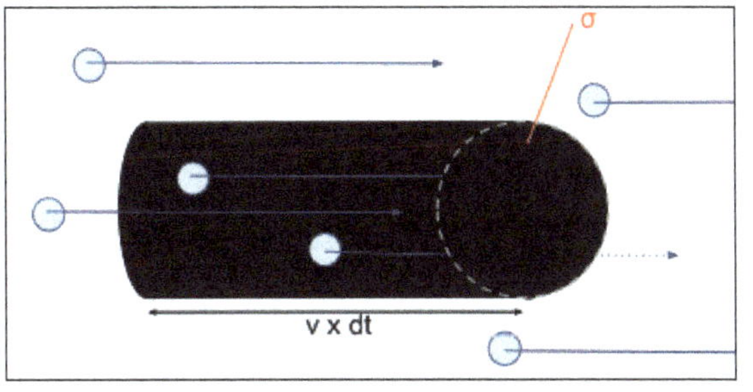

Interpretation of the reaction rate with the help of the cross section.

Noting n the number of particles per unit volume, there are $n\,V$ particles in the volume V, which will, per definition of V, undergo a reaction. Noting r the reaction rate onto one target, it gives:

$$r dt = nV = n\sigma v\, dt$$

It follows directly from the definition of the neutron flux $\Phi = n\,v$:

$$r = \sigma\Phi$$

Assuming that there is not one but N targets per unit volume, the reaction rate R per unit volume is:

$$R = Nr = N\Phi\sigma$$

Knowing that the typical nuclear radius r is of the order of 10^{-12} cm, the expected nuclear cross section is of the order of πr^2 or roughly 10^{-24} cm^2 (thus justifying the definition of the barn). However, if measured experimentally [$\sigma = R/(\Phi N)$], the experimental cross sections vary enormously. As an example, for slow neutrons absorbed by the (n, γ) reaction the cross section in some cases (xenon-135) is as much as 2,650,000 barns, while the cross sections for transmutations by gamma-ray absorption are in the neighborhood of 0.001 barn.

The "nuclear cross section" is consequently a purely conceptual quantity representing how big the nucleus should be to be consistent with this simple mechanical model.

Knowing that the typical nuclear radius r is of the order of 10^{-12} cm, the expected nuclear cross section is of the order of πr^2 or roughly 10^{-24} cm^2 (thus justifying the definition of the barn). However, if measured experimentally [$\sigma = R/(\Phi N)$], the experimental cross sections vary enormously. As an example, for slow neutrons absorbed by the (n, γ) reaction the cross section in some cases (xenon-135) is as much as 2,650,000 barns, while the cross sections for transmutations by gamma-ray absorption are in the neighborhood of 0.001 barn.

The "nuclear cross section" is consequently a purely conceptual quantity representing how big the nucleus should be to be consistent with this simple mechanical model.

Continuous versus Average Cross Section

Cross sections depend strongly on the incoming particle speed. In the case of a beam with multiple particle speeds, the reaction rate *R* is integrated over the whole range of energy:

$$R = \int_E N\Phi(E)\sigma(E)dE$$

where *σ(E)* is the continuous cross section, *Φ(E)* the differential flux and *N* the target atom density.

In order to obtain a formulation equivalent to the mono energetic case, an average cross section is defined:

$$\sigma = \frac{\int_E \Phi(E)\sigma(E)dE}{\int_E \Phi(E)dE} = \frac{\int_E \Phi(E)\sigma(E)dE}{\Phi}$$

where $\Phi = \int \Phi(E)\, dE$ is the integral flux.

Using the definition of the integral flux Φ and the average cross section σ, the same formulation as before is found:

$$R = N\Phi\sigma$$

Microscopic versus Macroscopic Cross Section

Up to now, the cross section referred corresponds to the microscopic cross section *σ*. However, it is possible to define the macroscopic cross section *Σ* which corresponds to the total "equivalent area" of all target particles per unit volume:

$$\Sigma = N\sigma$$

where *N* is the atomic density of the target.

Therefore, since the cross section can be expressed in cm^2 and the density in cm^{-3}, the macroscopic cross section is usually expressed in cm^{-1}. The reaction rate per unit volume *R* can be derived using only the neutron flux *Φ* and the macroscopic cross section *Σ*:

$$R = \Sigma\Phi$$

Mean Free Path

The mean free path *λ* of a random particle is the average length between two interactions. The total length *L* that non perturbed particles travel during a time interval *dt* in a volume *dV* is simply the product of the length *l* covered by each particle during this time with the number of particles *N* in this volume:

$$L = lN$$

Noting v the speed of the particles and n is the number of particles per unit volume:

$$l = v\,dt$$

$$N = n\,dV$$

It follows:

$$L = v\,dt\,n\,dV$$

Using the definition of the neutron flux Φ:

$$\Phi = nv$$

It follows:

$$L = \Phi\,dt\,dV$$

This average length L is however valid only for unperturbed particles. To account for the interactions, L is divided by the total number of reactions R to obtain the average length between each collision λ:

$$\lambda = \frac{L}{R} = \frac{\Phi\,dt\,dV}{R}$$

From Microscopic versus macroscopic cross section:

$$R = \Phi\Sigma\,dt\,dV$$

It follows:

$$\lambda = \frac{1}{\Sigma}$$

where λ is the mean free path and Σ is the macroscopic cross section.

Within Stars

Because lithium-8 and beryllium-12 form natural stopping points on the table of isotopes for hydrogen fusion, it is believed that all of the higher elements are formed in very hot stars where higher orders of fusion predominate. A star like the Sun produces energy by the fusion of simple H-1 into helium-4 through a series of reactions. It is believed that when the inner core exhausts its H-1 fuel, the Sun will contract, slightly increasing its core temperature until He-4 can fuse and become the main fuel supply. Pure He-4 fusion leads to Be-8, which decays back to 2 He-4; therefore the He-4 must fuse with isotopes either more or less massive than itself to result in an energy producing reaction. When He-4 fuses with H-2 or H-3, it forms stable isotopes Li-6 and Li-7 respectively. The higher order isotopes between Li-8 and C-12 are synthesized by similar reactions between hydrogen, helium, and lithium isotopes.

Typical Cross Sections

In the following, some cross sections which are of importance in a nuclear reactor are given. The thermal cross-section is averaged using a Maxwellian spectrum and the fast cross section is averaged using the uranium-235 fission spectrum.

Scattering		Thermal cross section (barn)			Fast cross section (barn)		
		Capture	Fission	Scattering	Capture	Fission	
Moderator	H-1	20	0.2	-	4	0.00004	-
	H-2	4	0.0003	-	3	0.000007	-
	C (natural)	5	0.002	-	2	0.00001	-
Structural materials, others	Au-197	8.2	98.7	-	4	0.08	-
	Zr-90	5	0.006	-	5	0.006	-
	Fe-56	10	2	-	20	0.003	-
	Cr-52	3	0.5	-	3	0.002	-
	Co-59	6	37.2	-	4	0.006	-
	Ni-58	20	3	-	3	0.008	-
	O-16	4	0.0001	-	3	0.00000003	-
Absorber	B-10	2	200	-	2	0.4	-
	Cd-113	100	30,000	-	4	0.05	-
	Xe-135	400,000	2,000,000	-	5	0.0008	-
	In-115	2	100	-	4	0.02	-
Fuel	U-235	10	99	583	4	0.09	1
	U-238	9	2	0.00002	5	0.07	0.3
	Pu-239	8	269	748	5	0.05	2

Nuclear Chain Reaction

A nuclear chain reaction occurs when one single nuclear reaction causes an average of one or more subsequent nuclear reactions, this leading to the possibility of a self-propagating series of these reactions. The specific nuclear reaction may be the fission of heavy isotopes (e.g., uranium-235, ^{235}U). The nuclear chain reaction releases several million times more energy per reaction than any chemical reaction.

Chemical chain reactions were first proposed by German chemist Max Bodenstein in 1913, and were reasonably well understood before nuclear chain reactions were proposed. It was understood that chemical chain reactions were responsible for exponentially increasing rates in reactions, such as produced in chemical explosions.

The concept of a nuclear chain reaction was reportedly first hypothesized by Hungarian scientist Leó Szilárd on September 12, 1933. Szilárd that morning had been reading in a London paper of an experiment in which protons from an accelerator had been used to split lithium-7 into alpha particles, and the fact that much greater amounts of energy were produced by the reaction than the proton supplied. Ernest Rutherford commented in the article that inefficiencies in the process precluded use of it for power generation. However, the neutron had been discovered in 1932, shortly before, as the product of a nuclear reaction. Szilárd, who had been trained as an engineer and physicist, put the two nuclear experimental results together in his mind and realized that if a nuclear reaction produced neutrons, which then caused further similar nuclear reactions, the process might be a self-perpetuating nuclear chain-reaction, spontaneously producing new isotopes and power without the need for protons or an accelerator. Szilárd, however, did not propose fission as the mechanism for his chain reaction, since the fission reaction was not yet discovered, or even suspected. Instead, Szilárd proposed using mixtures of lighter known isotopes which produced neutrons in copious amounts. He filed a patent for his idea of a simple nuclear reactor the following year.

In 1936, Szilárd attempted to create a chain reaction using beryllium and indium, but was unsuccessful. Nuclear fission was discovered and proved by Otto Hahn and Fritz Strassmann in December 1938. A few months later, Frédéric Joliot, H. Von Halban and L. Kowarski in Paris searched for, and discovered, neutron multiplication in uranium, proving that a nuclear chain reaction by this mechanism was indeed possible.

On May 4, 1939 Joliot, Halban et Kowarski filed three patents. The first two described power production from a nuclear chain reaction, the last one called "*Perfectionnement aux charges explosives*" was the first patent for the atomic bomb and is filed as patent n°445686 by the Caisse nationale de Recherche Scientifique.

In parallel, Szilárd and Enrico Fermi in New York made the same analysis. This discovery prompted the letter from Szilárd and signed by Albert Einstein to President Franklin D. Roosevelt, warning of the possibility that Nazi Germany might be attempting to build an atomic bomb.

On December 2, 1942, a team led by Enrico Fermi (and including Szilárd) produced the first artificial self-sustaining nuclear chain reaction with the Chicago Pile-1 (CP-1) experimental reactor in a racquets court below the bleachers of Stagg Field at the University of Chicago. Fermi's experiments at the University of Chicago were part of Arthur H. Compton's Metallurgical Laboratory of the Manhattan Project; the lab was later renamed Argonne National Laboratory, and tasked with conducting research in harnessing fission for nuclear energy.

In 1956, Paul Kuroda of the University of Arkansas postulated that a natural fission reactor may have once existed. Since nuclear chain reactions may only require natural materials (such as water and uranium, if the uranium has sufficient amounts of U-235), it was possible to have these chain reactions occur in the distant past when uranium-235 concentrations were higher than today, and where there was the right combination of materials within the Earth's crust. Kuroda's prediction was verified with the discovery of evidence of natural self-sustaining nuclear chain reactions in the past at Oklo in Gabon, Africa, in September 1972.

Fission Chain Reaction

Fission chain reactions occur because of interactions between neutrons and fissile isotopes (such as ^{235}U). The chain reaction requires both the release of neutrons from fissile isotopes undergoing nuclear fission and the subsequent absorption of some of these neutrons in fissile isotopes. When an atom undergoes nuclear fission, a few neutrons (the exact number depends on uncontrollable and unmeasurable factors; the expected number depends on several factors, usually between 2.5

and 3.0) are ejected from the reaction. These free neutrons will then interact with the surrounding medium, and if more fissile fuel is present, some may be absorbed and cause more fissions. Thus, the cycle repeats to give a reaction that is self-sustaining.

Nuclear power plants operate by precisely controlling the rate at which nuclear reactions occur, on the other hand, nuclear weapons are specifically engineered to produce a reaction that is so fast and intense it cannot be controlled after it has started. When properly designed, this uncontrolled reaction will lead to an explosive energy release.

Nuclear Fission Fuel

Nuclear weapons employ high quality, highly enriched fuel exceeding the critical size and geometry (critical mass) necessary in order to obtain an explosive chain reaction. The fuel for energy purposes, such as in a nuclear fission reactor, is very different, usually consisting of a low-enriched oxide material (e.g. UO_2).

Fission Reaction Products

When a fissile atom undergoes nuclear fission, it breaks into two or more fission fragments. Also, several free neutrons, gamma rays, and neutrinos are emitted, and a large amount of energy is released. The sum of the rest masses of the fission fragments and ejected neutrons is less than the sum of the rest masses of the original atom and incident neutron (of course the fission fragments are not at rest). The mass difference is accounted for in the release of energy according to the equation $E=\Delta mc^2$:

$$\text{mass of released energy} = \frac{E}{c^2} = m_{\text{original}} - m_{\text{final}}$$

Due to the extremely large value of the speed of light, c, a small decrease in mass is associated with a tremendous release of active energy (for example, the kinetic energy of the fission fragments). This energy (in the form of radiation and heat) carries the missing mass, when it leaves the reaction system (total mass, like total energy, is always conserved). While typical chemical reactions release energies on the order of a few eVs (e.g. the binding energy of the electron to hydrogen is 13.6 eV), nuclear fission reactions typically release energies on the order of hundreds of millions of eVs.

Two typical fission reactions are shown below with average values of energy released and number of neutrons ejected:

$$^{235}U + neutron \rightarrow \text{fission fragments} + 2.4 \text{ neutrons} + 192.9 \text{ MeV}$$

$$^{239}Pu + neutron \rightarrow \text{fission fragments} + 2.9 \text{ neutrons} + 198.5 \text{ MeV}$$

Note that these equations are for fissions caused by slow-moving (thermal) neutrons. The average energy released and number of neutrons ejected is a function of the incident neutron speed. Also, note that these equations exclude energy from neutrinos since these subatomic particles are extremely non-reactive and, therefore, rarely deposit their energy in the system.

Timescales of Nuclear Chain Reactions

Prompt Neutron Lifetime

The prompt neutron lifetime, *l*, is the average time between the emission of neutrons and either their absorption in the system or their escape from the system. The neutrons that occur directly from fission are called "prompt neutrons," and the ones that are a result of radioactive decay of fission fragments are called "delayed neutrons". The term lifetime is used because the emission of a neutron is often considered its "birth," and the subsequent absorption is considered its "death". For thermal (slow-neutron) fission reactors, the typical prompt neutron lifetime is on the order of 10^{-4} seconds, and for fast fission reactors, the prompt neutron lifetime is on the order of 10^{-7} seconds. These extremely short lifetimes mean that in 1 second, 10,000 to 10,000,000 neutron lifetimes can pass. The *average* (also referred to as the *adjoint unweighted*) prompt neutron lifetime takes into account all prompt neutrons regardless of their importance in the reactor core; the *effective* prompt neutron lifetime (referred to as the *adjoint weighted* over space, energy, and angle) refers to a neutron with average importance.

Mean Generation Time

The mean generation time, Λ, is the average time from a neutron emission to a capture that results in fission. The mean generation time is different from the prompt neutron lifetime because the mean generation time only includes neutron absorptions that lead to fission reactions (not other absorption reactions). The two times are related by the following formula:

$$\Lambda = \frac{l}{k}$$

In this formula, k is the effective neutron multiplication factor.

Effective Neutron Multiplication Factor

The six factor formula effective neutron multiplication factor, *k*, is the average number of neutrons from one fission that cause another fission. The remaining neutrons either are absorbed in non-fission reactions or leave the system without being absorbed. The value of *k* determines how a nuclear chain reaction proceeds:

- $k < 1$ (subcriticality): The system cannot sustain a chain reaction, and any beginning of a chain reaction dies out over time. For every fission that is induced in the system, an average *total* of $1/(1 - k)$ fissions occur.
- $k = 1$ (criticality): Every fission causes an average of one more fission, leading to a fission (and power) level that is constant. Nuclear power plants operate with $k = 1$ unless the power level is being increased or decreased.
- $k > 1$ (supercriticality): For every fission in the material, it is likely that there will be "*k*" fissions after the next *mean generation time* (Λ). The result is that the number of fission reactions increases exponentially, according to the equation $e^{(k-1)t/\Lambda}$, where t is the elapsed time. Nuclear weapons are designed to operate under this state. There are two subdivisions of supercriticality: prompt and delayed.

When describing kinetics and dynamics of nuclear reactors, and also in the practice of reactor operation, the concept of reactivity is used, which characterizes the deflection of reactor from the critical state. $\rho=(k-1)/k$. In Hour is a unit of reactivity of a nuclear reactor.

In a nuclear reactor, *k* will actually oscillate from slightly less than 1 to slightly more than 1, due primarily to thermal effects (as more power is produced, the fuel rods warm and thus expand, lowering their capture ratio, and thus driving *k* lower). This leaves the average value of *k* at exactly 1. Delayed neutrons play an important role in the timing of these oscillations.

In an infinite medium, the multiplication factor may be described by the four factor formula; in a non-infinite medium, the multiplication factor may be described by the six factor formula.

Prompt and Delayed Supercriticality

Not all neutrons are emitted as a direct product of fission; some are instead due to the radioactive decay of some of the fission fragments. The neutrons that occur directly from fission are called "prompt neutrons," and the ones that are a result of radioactive decay of fission fragments are called "delayed neutrons". The fraction of neutrons that are delayed is called β, and this fraction is typically less than 1% of all the neutrons in the chain reaction.

The delayed neutrons allow a nuclear reactor to respond several orders of magnitude more slowly than just prompt neutrons would alone. Without delayed neutrons, changes in reaction rates in nuclear reactors would occur at speeds that are too fast for humans to control.

The region of supercriticality between $k = 1$ and $k = 1/(1-\beta)$ is known as delayed supercriticality (or delayed criticality). It is in this region that all nuclear power reactors operate. The region of supercriticality for $k > 1/(1-\beta)$ is known as prompt supercriticality (or prompt criticality), which is the region in which nuclear weapons operate.

The change in k needed to go from critical to prompt critical is defined as a dollar.

Nuclear Weapons Application of Neutron Multiplication

Nuclear fission weapons require a mass of fissile fuel that is prompt supercritical. For a given mass of fissile material the value of k can be increased by increasing the density. Since the probability per distance traveled for a neutron to collide with a nucleus is proportional to the material density, increasing the density of a fissile material can increase k. This concept is utilized in the implosion method for nuclear weapons. In these devices, the nuclear chain reaction begins after increasing the density of the fissile material with a conventional explosive.

In the gun-type fission weapon two subcritical pieces of fuel are rapidly brought together. The value of k for a combination of two masses is always greater than that of its components. The magnitude of the difference depends on distance, as well as the physical orientation.

The value of k can also be increased by using a neutron reflector surrounding the fissile material.

Once the mass of fuel is prompt supercritical, the power increases exponentially. However, the exponential power increase cannot continue for long since k decreases when the amount of fission material that is left decreases (i.e. it is consumed by fissions). Also, the geometry and density are expected to change during detonation since the remaining fission material is torn apart from the explosion.

Predetonation

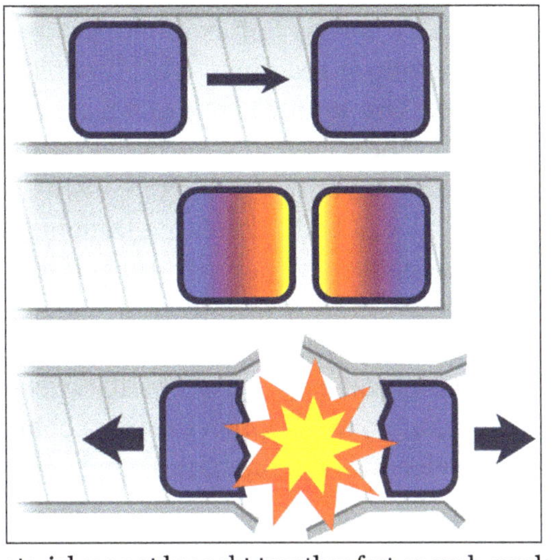

If two pieces of subcritical material are not brought together fast enough, nuclear predetonation can occur, whereby a smaller explosion than expected will blow the bulk of the material apart.

Detonation of a nuclear weapon involves bringing fissile material into its optimal supercritical state very rapidly. During part of this process, the assembly is supercritical, but not yet in an optimal state for a chain reaction. Free neutrons, in particular from spontaneous fissions, can cause the device to undergo a preliminary chain reaction that destroys the fissile material before it is ready to produce a large explosion, which is known as predetonation.

To keep the probability of predetonation low, the duration of the non-optimal assembly period is minimized and fissile and other materials are used which have low spontaneous fission rates. In fact, the combination of materials has to be such that it is unlikely that there is even a single spontaneous fission during the period of supercritical assembly. In particular, the gun method cannot be used with plutonium.

Nuclear Power Plants and Control of Chain Reactions

Chain reactions naturally give rise to reaction rates that grow (or shrink) exponentially, whereas a nuclear power reactor needs to be able to hold the reaction rate reasonably constant. To maintain this control, the chain reaction criticality must have a slow enough time-scale to permit intervention by additional effects (e.g., mechanical control rods or thermal expansion). Consequently, all nuclear power reactors (even fast-neutron reactors) rely on delayed neutrons for their criticality. An operating nuclear power reactor fluctuates between being slightly subcritical and slightly delayed-supercritical, but must always remain below prompt-critical.

It is impossible for a nuclear power plant to undergo a nuclear chain reaction that results in an explosion of power comparable with a nuclear weapon, but even low-powered explosions due to uncontrolled chain reactions, that would be considered "fizzles" in a bomb, may still cause considerable damage and meltdown in a reactor. For example, the Chernobyl disaster involved a runaway chain reaction but the result was a low-powered steam explosion from the relatively small release of heat, as compared with a bomb. However, the reactor complex was destroyed by the heat, as well as by ordinary burning of the graphite exposed to air. Such steam explosions would be typical of the very diffuse assembly of materials in a nuclear reactor, even under the worst conditions.

In addition, other steps can be taken for safety. For example, power plants licensed in the United States require a negative void coefficient of reactivity (this means that if water is removed from the reactor core, the nuclear reaction will tend to shut down, not increase). This eliminates the possibility of the type of accident that occurred at Chernobyl (which was due to a positive void coefficient). However, nuclear reactors are still capable of causing smaller explosions even after complete shutdown, such as was the case of the Fukushima Daiichi nuclear disaster. In such cases, residual decay heat from the core may cause high temperatures if there is loss of coolant flow, even a day after the chain reaction has been shut down. This may cause a chemical reaction between water and fuel that produces hydrogen gas which can explode after mixing with air, with severe contamination consequences, since fuel rod material may still be exposed to the atmosphere from this process. However, such explosions do not happen during a chain reaction, but rather as a result of energy from radioactive beta decay, after the fission chain reaction has been stopped.

Nuclear Fission

Nuclear fission is the subdivision of a heavy atomic nucleus, such as that of uranium or plutonium, into two fragments of roughly equal mass. The process is accompanied by the release of a large amount of energy.

In nuclear fission the nucleus of an atom breaks up into two lighter nuclei. The process may take place spontaneously in some cases or may be induced by the excitation of the nucleus with a variety of particles (e.g., neutrons, protons, deuterons, or alpha particles) or with electromagnetic radiation in the form of gamma rays. In the fission process, a large quantity of energy is released, radioactive products are formed, and several neutrons are emitted. These neutrons can induce fission in a nearby nucleus of fissionable material and release more neutrons that can repeat the sequence, causing a chain reaction in which a large number of nuclei undergo fission and an enormous amount of energy is released. If controlled in a nuclear reactor, such a chain reaction can provide power for society's benefit. If uncontrolled, as in the case of the so-called atomic bomb, it can lead to an explosion of awesome destructive force.

The discovery of nuclear fission has opened a new era—the "Atomic Age." The potential of nuclear fission for good or evil and the risk/benefit ratio of its applications have not only provided the basis of many sociological, political, economic, and scientific advances but grave concerns as well. Even from a purely scientific perspective, the process of nuclear fission has given rise to many puzzles and complexities, and a complete theoretical explanation is still not at hand.

The term fission was first used by the German physicists Lise Meitner and Otto Frisch in 1939 to describe the disintegration of a heavy nucleus into two lighter nuclei of approximately equal size. The conclusion that such an unusual nuclear reaction can in fact occur was the culmination of a truly dramatic episode in the history of science, and it set in motion an extremely intense and productive period of investigation.

The story of the discovery of nuclear fission actually began with the discovery of the neutron in 1932 by James Chadwick in England. Shortly thereafter Enrico Fermi and his associates in Italy undertook an extensive investigation of the nuclear reactions produced by the bombardment of various elements with this uncharged particle. In particular, these workers observed that at least four different radioactive species resulted from the bombardment of uranium with slow neutrons. These newly discovered species emitted beta particles and were thought to be isotopes of unstable "transuranium elements" of atomic numbers 93, 94, and perhaps higher. There was, of course, intense interest in examining the properties of these elements, and many radiochemists participated in the studies. The results of these investigations, however, were extremely perplexing, and confusion persisted until 1939 when Otto Hahn and Fritz Strassmann in Germany, following a clue provided by Irène Joliot-Curie and Pavle Savić in France, proved definitely that the so-called transuranic elements were in fact radioisotopes of barium, lanthanum, and other elements in the middle of the periodic table.

That lighter elements could be formed by bombarding heavy nuclei with neutrons had been suggested earlier, but the idea was not given serious consideration because it entailed such a broad departure from the accepted views of nuclear physics and was unsupported by clear chemical evidence. Armed with the unequivocal results of Hahn and Strassmann, however, Meitner and Frisch invoked the recently formulated liquid-drop model of the nucleus to give a qualitative theoretical interpretation of the fission process and called attention to the large energy release that should accompany it. There was almost immediate confirmation of this reaction in dozens of laboratories throughout the world, and within a year more than 100 papers describing most of the important features of the process were published. These experiments confirmed the formation of extremely energetic heavy particles and extended the chemical identification of the products.

The chemical evidence that was so vital in leading Hahn and Strassmann to the discovery of nuclear fission was obtained by the application of carrier and tracer techniques. Since invisible amounts of the radioactive species were formed, their chemical identity had to be deduced from the manner in which they followed known carrier elements, present in macroscopic quantity, through various chemical operations. Known radioactive species were also added as tracers and their behavior was compared with that of the unknown species to aid in the identification of the latter. Over the years, these radiochemical techniques have been used to isolate and identify some 34 elements from zinc (atomic number 30) to gadolinium (atomic number 64) that are formed as fission products. The wide range of radioactivities produced in fission makes this reaction a rich source of tracers for chemical, biologic, and industrial use.

Although the early experiments involved the fission of ordinary uranium with slow neutrons, it was rapidly established that the rare isotope uranium-235 was responsible for this phenomenon. The more abundant isotope uranium-238 could be made to undergo fission only by fast neutrons with energy exceeding 1 MeV. The nuclei of other heavy elements, such as thorium and protactinium, also were shown to be fissionable with fast neutrons; and other particles, such as fast protons, deuterons, and alphas, along with gamma rays, proved to be effective in inducing the reaction.

In 1939, Frédéric Joliot-Curie, Hans von Halban, and Lew Kowarski found that several neutrons were emitted in the fission of uranium-235, and this discovery led to the possibility of a self-sustaining chain reaction. Fermi and his coworkers recognized the enormous potential of such a reaction if it could be controlled. On Dec. 2, 1942, they succeeded in doing so, operating the world's first nuclear reactor. Known as a "pile," this device consisted of an array of uranium and graphite blocks and was built on the campus of the University of Chicago.

The secret Manhattan Project, established not long after the United States entered World War II, developed the atomic bomb. Once the war had ended, efforts were made to develop new reactor types for large-scale power generation, giving birth to the nuclear power industry.

Fundamentals of the Fission Process

Structure and Stability of Nuclear Matter

The fission process may be best understood through a consideration of the structure and stability of nuclear matter. Nuclei consist of nucleons (neutrons and protons), the total number of which is equal to the mass number of the nucleus. The actual mass of a nucleus is always less than the sum of the masses of the free neutrons and protons that constitute it, the difference being the mass equivalent of the energy of formation of the nucleus from its constituents. The conversion of mass to energy follows Einstein's equation, $E = mc^2$, where E is the energy equivalent of a mass, m, and c is the velocity of light. This difference is known as the mass defect and is a measure of the total binding energy (and, hence, the stability) of the nucleus. This binding energy is released during the formation of a nucleus from its constituent nucleons and would have to be supplied to the nucleus to decompose it into its individual nucleon components.

A curve illustrating the average binding energy per nucleon as a function of the nuclear mass number is shown in figure . The largest binding energy (highest stability) occurs near mass number 56—the mass region of the element iron. Figure indicates that any nucleus heavier than mass number 56 would become a more stable system by breaking into lighter nuclei of higher binding energy, the difference in binding energy being released in the process. It should be noted that nuclei lighter than mass number 56 can gain in stability by fusing to produce a heavier nucleus of greater mass defect—again, with the release of the energy equivalent of the mass difference. It is the fusion of the lightest nuclei that provides the energy released by the Sun and constitutes the basis of the hydrogen, or fusion, bomb. Efforts to harness fusion reaction for power production have been actively pursued.

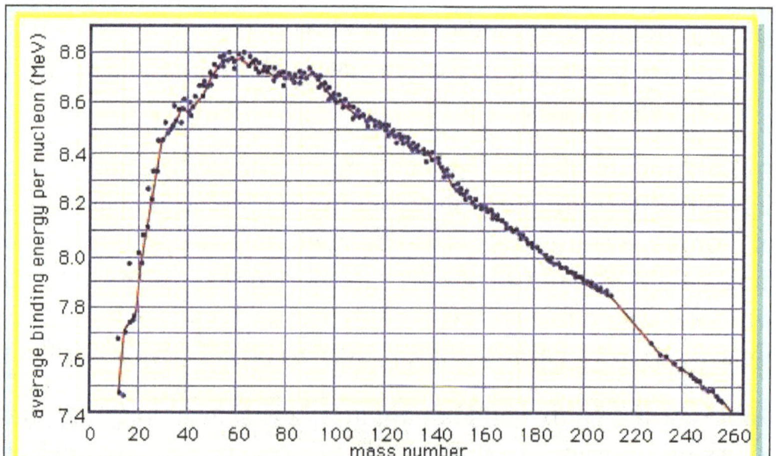

The average binding energy per nucleon as a function of the mass number, A. The line connects the odd-A points.

On the basis of energy considerations alone, figure would indicate that all matter should seek its most stable configuration, becoming nuclei of mass number near 56. However, this does not happen, because barriers to such a spontaneous conversion are provided by other factors. A good qualitative understanding of the nucleus is achieved by treating it as analogous to a uniformly charged liquid drop. The strong attractive nuclear force between pairs of nucleons is of short range and acts only between the closest neighbours. Since nucleons near the surface of the drop have fewer close neighbours than those in the interior, a surface tension is developed, and the nuclear drop assumes a spherical shape in order to minimize this surface energy. (The smallest surface area enclosing a given volume is provided by a sphere.) The protons in the nucleus exert a long-range repulsive (Coulomb) force on each other because of their positive charge. As the number of nucleons in a nucleus increases beyond about 40, the number of protons must be diluted with an excess of neutrons to maintain relative stability.

If the nucleus is excited by some stimulus and begins to oscillate (i.e., deform from its spherical shape), the surface forces will increase and tend to restore it to a sphere, where the surface tension is at a minimum. On the other hand, the Coulomb repulsion decreases as the drop deforms and the protons are positioned farther apart. These opposing tendencies set up a barrier in the potential energy of the system, as indicated in figure.

The curve in figure rises initially with elongation, since the strong, short-range nuclear force that gives rise to the surface tension increases. The Coulomb repulsion between protons decreases faster with elongation than the surface tension increases, and the two are in balance at point B, which represents the height of the barrier to fission. (This point is called the "saddle point" because, in a three-dimensional view of the potential energy surface, the shape of the pass over the barrier resembles a saddle.) Beyond point B, the Coulomb repulsion between the protons drives the nucleus into further elongation until at some point, S (the scission point), the nucleus breaks in two. Qualitatively, at least, the fission process is thus seen to be a consequence of the Coulomb repulsion between protons.

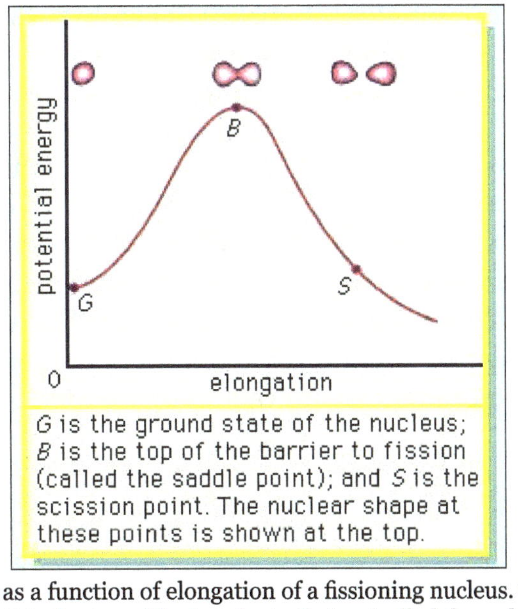

The potential energy as a function of elongation of a fissioning nucleus. G is the ground state of the nucleus; B is the top of the barrier to fission (called the saddle point); and S is the scission point. The nuclear shape at these points is shown at the top.

Induced Fission

The height and shape of the fission barrier are dependent on the particular nucleus being considered. Fission can be induced by exciting the nucleus to an energy equal to or greater than that of the barrier. This can be done by gamma-ray excitation (photofission) or through excitation of the nucleus by the capture of a neutron, proton, or other particle (particle-induced fission). The binding energy of a particular nucleon to a nucleus will depend on—in addition to the factors considered above—the odd–even character of the nucleus. Thus, if a neutron is added to a nucleus having an odd number of neutrons, an even number of neutrons will result, and the binding energy will be greater than for the addition of a neutron that makes the total number of neutrons odd. This "pairing energy" accounts in part for the difference in behavior of nuclides in which fission can be induced with slow (low-energy) neutrons and those that require fast (higher-energy) neutrons. Although the heavy elements are unstable with respect to fission, the reaction takes place to an appreciable extent only if sufficient energy of activation is available to surmount the fission barrier. Most nuclei that are fissionable with slow neutrons contain an odd number of neutrons (e.g., uranium-233, uranium-235, or plutonium-239), whereas most of those requiring fast neutrons (e.g., thorium-232 or uranium-238) have an even number. The addition of a neutron in the former case liberates sufficient binding energy to induce fission. In the latter case, the binding energy is less and may be insufficient to surmount the barrier and induce fission. Additional energy must then be supplied in the form of the kinetic energy of the incident neutron. (In the case of thorium-232 or uranium-238, a neutron having about 1 MeV of kinetic energy is required).

Spontaneous Fission

The laws of quantum mechanics deal with the probability of a system such as a nucleus or an atom being in any of its possible states or configurations at any given time. A fissionable system (uranium-238, for example) in its ground state (i.e., at its lowest excitation energy and with an elongation small enough that it is confined inside the fission barrier) has a small but finite probability of being in the energetically favoured configuration of two fission fragments. In effect, when this occurs, the system has penetrated the barrier by the process of quantum mechanical tunneling. This process is called spontaneous fission because it does not involve any outside influences. In the case of uranium-238, the process has a very low probability, requiring more than 10^{15} years for half of the material to be transformed (its so-called half-life) by this reaction. On the other hand, the probability for spontaneous fission increases dramatically for the heaviest nuclides known and becomes the dominant mode of decay for some—those having half-lives of only fractions of a second. In fact, spontaneous fission becomes the limiting factor that may prevent the formation of still heavier (super-heavy) nuclei.

The Stages of Fission

A pictorial representation of the sequence of events in the fission of a heavy nucleus is given in figure. The approximate time elapse between stages of the process is indicated at the bottom of the figure.

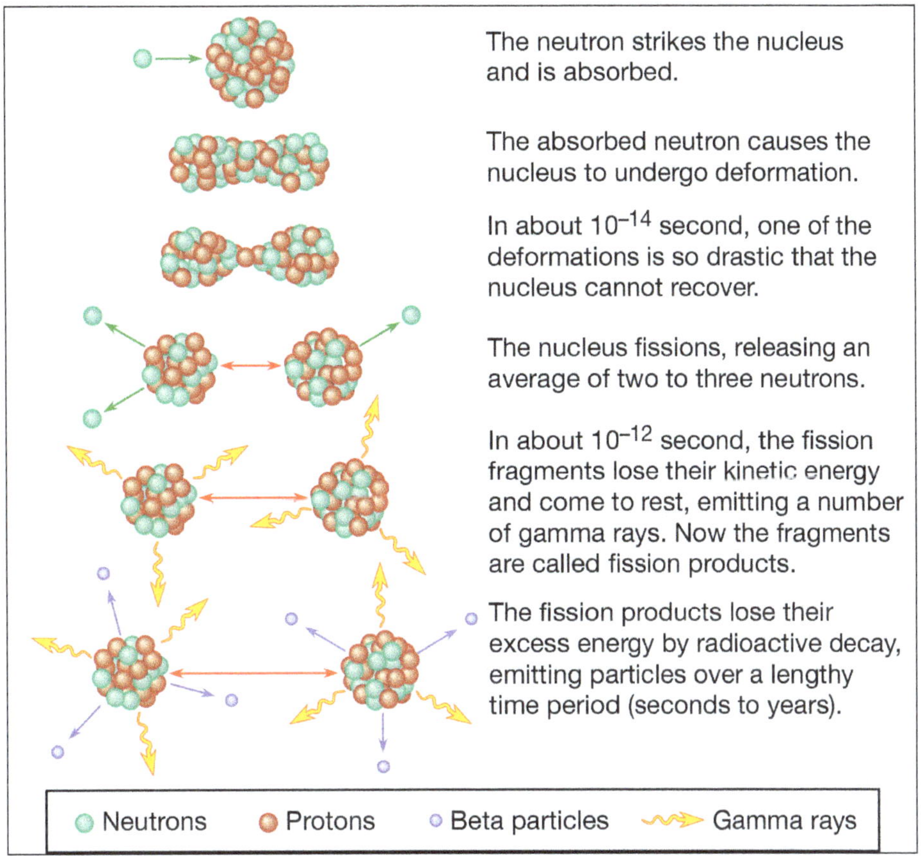

Sequence of events in the fission of a uranium nucleus by a neutron.

The Phenomenology of Fission

When a heavy nucleus undergoes fission, a variety of fragment pairs may be formed, depending on the distribution of neutrons and protons between the fragments. This leads to probability distribution of both mass and nuclear charge for the fragments. The probability of formation of a particular fragment is called its fission yield and is expressed as the percentage of fissions leading to it.

The separated fragments experience a large Coulomb repulsion due to their nuclear charges, and they recoil from each other with kinetic energies determined by the fragment charges and the distance between the charge centres at the time of scission. Variations in these parameters lead to a distribution of kinetic energies, even for the same mass split.

The initial velocities of the recoiling fragments are too fast for the outer (atomic) electrons of the fissioning atom to keep pace, and many of them are stripped away. Thus, the nuclear charge of the fragment is not fully neutralized by the atomic electrons, and the fission fragments fly apart as highly charged atoms. As the nucleus of the fragment adjusts from its deformed shape to a more stable configuration, the deformation energy (i.e., the energy required to deform it) is recovered and converted into internal excitation energy, and neutrons and prompt gamma rays (an energetic form of electromagnetic radiation given off nearly coincident with the fission event) may be evaporated from the moving fragment. The fast-moving, highly charged atom collides with the atoms of the medium through which it is moving, and its kinetic energy is transferred to ionization and heating of the medium as it slows down and comes to rest. The range of fission fragments in air is only a few centimetres.

During the slowing-down process, the charged atom picks up electrons from the medium and becomes neutral by the time it stops. At this stage in the sequence of events, the atom produced is called a fission product to distinguish it from the initial fission fragment formed at scission. Since a few neutrons may have been lost in the transition from fission fragment to fission product, the two may not have the same mass number. The fission product is still not a stable species but is radioactive, and it finally reaches stability by undergoing a series of beta decays, which may vary over a time scale of fractions of a second to many years. The beta emission consists of electrons and antineutrinos, often accompanied by gamma rays and X-rays.

The distributions in mass, charge, and kinetic energy of the fragments have been found to be dependent on the fissioning species as well as on the excitation energy at which the fission act occurs. Many other aspects of fission have been observed, adding to the extensive phenomenology of the process and providing an intriguing set of problems for interpretation. These include the systematics of fission cross sections (a measure of the probability for fission to occur); the variation of the number of prompt neutrons emitted as a function of the fissioning species and the particular fragment mass split; the angular distribution of the fragments with respect to the direction of the beam of particles inducing fission; the systematics of spontaneous fission half-lives; the occurrence of spontaneous fission isomers (excited states of the nucleus); the emission of light particles (hydrogen-3, helium-3, helium-4, etc.) in small but significant numbers in some fission events; the presence of delayed neutron emitters among the fission products; the time scale on which the various stages of the process take place; and the distribution of the energy release in fission among the particles and radiations produced.

Fission Fragment Mass Distributions

The distribution of the fragment masses formed in fission is one of the most striking features of the process. It is dependent on the mass of the fissioning nucleus and the excitation energy at which the fission occurs. At low excitation energy, the fission of such nuclides as uranium-235 or plutonium-239 is asymmetric; i.e., the fragments are formed in a two-humped probability (or yield) distribution favouring an unequal division in mass. This is illustrated in figure. As will be noted, the light group of fragment masses shifts to higher mass numbers as the mass of the fissioning nucleus increases, whereas the position of the heavy group remains nearly stationary. As the excitation energy of the fission increases, the probability for a symmetric mass split increases, while that for asymmetric division decreases. Thus, the valley between the two peaks increases in probability (yield of formation), and at high excitations the mass distribution becomes single-humped, with the maximum yield at symmetry. Radium isotopes show interesting triple-humped mass distributions, and nuclides lighter than radium show a single-humped, symmetric mass distribution. (These nuclides, however, require a relatively high activation energy to undergo fission.) For very heavy nuclei in the region of fermium-260, the mass-yield curve becomes symmetric (single-humped) even for spontaneous fission, and the kinetic energies of the fragments are unusually high. An understanding of these mass distributions has been one of the major puzzles of fission, and a complete theoretical interpretation is still lacking, albeit much progress has been made.

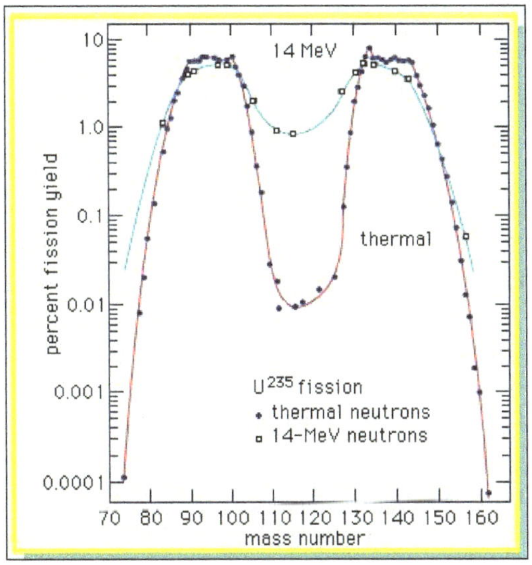

Mass distribution dependence on the energy excitation in the fission of uranium-235.
At still higher energies, the curve becomes single-humped, with a maximum yield for symmetric mass splits.

Fission Decay Chains and Charge Distribution

In order to maintain stability, the neutron-to-proton (n/p) ratio in nuclei must increase with increasing proton number. The ratio remains at unity up to the element calcium, with 20 protons. It then gradually increases until it reaches a value of about 1.5 for the heaviest elements. When a heavy nucleus fissions, a few neutrons are emitted; however, this still leaves too high an n/p ratio in the fission fragments to be consistent with stability for them. They undergo radioactive decay and reach stability by successive conversions of neutrons to protons with the emission of a negative electron (called a beta particle, β^-) and an antineutrino. The mass number of the nucleus remains the same, but the nuclear charge (atomic number) increases by one, and a new element is formed for each such conversion. The successive beta decays constitute an isobaric, fission-product decay chain for each mass number. The half-lives for the decay of the radioactive species generally increase as they approach the stable isobar of the chain. (Species of the same element characterized by the same nuclear charge, Z [number of protons], but differing in their number of neutrons [and therefore in mass number A] are called isotopes. Species that have the same mass number, A, but differ in Z are known as isobars.)

For a typical mass split in the neutron-induced fission of uranium-235, the complementary fission-product masses of 93 and 141 may be formed following the emission of two neutrons from the initial fragments. The division of charge (i.e., protons) between the fragments represents an important parameter in the fission process. Thus, for the mass numbers 93 and 141, the following isobaric fission-product decay chains would be formed (the half-lives for the beta-decay processes are indicated above the arrows):

$$^{93}_{36}\mathrm{Kr} \xrightarrow[\beta^-]{1.3\text{ sec}} {}^{93}_{37}\mathrm{Rb} \xrightarrow[\beta^-]{6\text{ sec}} {}^{93}_{38}\mathrm{Sr} \xrightarrow[\beta^-]{7.5\text{ min}} {}^{93}_{39}\mathrm{Y} \xrightarrow[\beta^-]{10\text{ hr}} {}^{93}_{40}\mathrm{Zr} \xrightarrow{1.5\times10^{6}\text{ yr}} {}^{93m}_{41}\mathrm{Nb} \xrightarrow{13\text{ yr}} {}^{93}_{41}\mathrm{Nb}\ \text{(stable)}$$

$$^{93}_{40}\mathrm{Zr} \xrightarrow{\beta^-} {}^{93}_{41}\mathrm{Nb}\ \text{(stable)}$$

$$^{141}_{54}\mathrm{Xe} \xrightarrow[\beta^-]{1.7\text{ sec}} {}^{141}_{55}\mathrm{Cs} \xrightarrow[\beta^-]{25\text{ sec}} {}^{141}_{56}\mathrm{Ba} \xrightarrow[\beta^-]{18\text{ min}} {}^{141}_{57}\mathrm{La} \xrightarrow[\beta^-]{3.9\text{ hr}} {}^{141}_{58}\mathrm{Ce} \xrightarrow[\beta^-]{32.5\text{ d}} {}^{141}_{59}\mathrm{Pr}\ \text{(stable)}$$

(The left subscript on the element symbol denotes Z, while the superscript denotes A.) The 92 protons of the uranium nucleus must be conserved, and complementary fission-product pairs—such as krypton-36 with barium-56, rubidium-37 with cesium-55, or strontium-38 with xenon-54—would be possible.

The percentage of fissions in which a particular nuclide is formed as a primary fission product (i.e., as the direct descendant of an initial fragment following its de-excitation) is called the independent yield of that product. The total yield for any nuclide in the isobaric decay chain is the sum of its independent yield and the independent yields of all of its precursors in the chain. The total yield for the entire chain is called the cumulative yield for that mass number.

Extensive radiochemical investigations have suggested that the most probable charge division is one that is displaced from stability about the same distance in both chains. This empirical observation is called the equal charge displacement (ECD) hypothesis, and it has been confirmed by several physical measurements. In the above example the ECD would predict the most probable charges at about rubidium-37 and cesium-55. A strong shell effect modifies the ECD expectations for fragments having 50 protons. The dispersion of the charge formation probability about the most probable charge (Z_p) is rather narrow and approximately Gaussian in shape and is nearly independent of the mass split as well as of the fissioning species. The most probable charge for an isobaric chain is a useful concept in the description of the charge dispersion, and it need not have an integral value. As the energy of fission increases, the charge division tends toward maintaining the n/p ratio in the fragments the same as that in the fissioning nucleus. This is referred to as an unchanged charge distribution.

Prompt Neutrons in Fission

The average number of neutrons emitted per fission (represented by the symbol $\bar{\nu}$) varies with the fissioning nucleus. It is about 2.0 for the spontaneous fission of uranium-238 and 4.0 for that of fermium-257. In the thermal-neutron induced fission of uranium-235, $\bar{\nu}$= 2.4. The actual number of neutrons emitted, however, varies with each fission event, depending on the mass split. Although there is still controversy regarding the number of neutrons emitted at the instant of scission, it is generally agreed that most of the neutrons are given off by the recoiling fission fragments soon after scission occurs. The number of neutrons emitted from each fragment depends on the amount of energy the fragment possesses. The energy can be in the form of internal excitation (heat) energy or stored as energy of deformation of the fragment to be released when the fragment returns to its stable equilibrium shape.

Delayed Neutrons in Fission

A few of the fission products have beta-decay energies that exceed the binding energy of a neutron in the daughter nucleus. This is likely to happen when the daughter nucleus contains one or two neutrons more than a closed shell of 50 or 82 neutrons, since these "extra" neutrons are more loosely bound. The beta decay of the precursor may take place to an excited state of the daughter from which a neutron is emitted. The neutron emission is "delayed" by the beta-decay half-life of the precursor. Six such delayed neutron emitters have been identified, with half-lives varying from about 0.5 to 56 seconds. The yield of the delayed neutrons is only about 1 percent of that of the prompt neutrons, but they are very important for the control of the chain reaction in a nuclear reactor.

Energy Release in Fission

The total energy release in a fission event may be calculated from the difference in the rest masses of the reactants (e.g., ^{235}U + n) and the final stable products (e.g., ^{93}Nb + ^{141}Pr + 2n). The energy equivalent of this mass difference is given by the Einstein relation, $E = mc^2$. The total energy release depends on the mass split, but a typical fission event would have the total energy release distributed approximately as follows for the major components in the thermal neutron-induced fission of uranium-235:

Energy component	Number per fission	Total energy
Kinetic energy of fission fragments	2	170MeV
Kinetic energy of prompt neutrons	2.5	5
Binding energy from capture of prompt neutrons	2.5	~12
Prompt gamma rays	8	8
Total=195 MeV		

(The energy release from the capture of the prompt neutrons depends on how they are finally stopped, and some will escape the core of a nuclear reactor).

This energy is released on a time scale of about 10^{-12} second and is called the prompt energy release. It is largely converted to heat within an operating reactor and is used for power generation. Also, there is a delayed release of energy from the radioactive decay of the fission products varying in half-life from fractions of a second to many years. The shorter-lived species decay in the reactor, and their energy adds to the heat generated; however, the longer-lived species remain radioactive and pose a problem in the handling and disposition of the reactor fuel elements when they need to be replaced. The antineutrinos that accompany the beta decay of the fission products are unreactive, and their kinetic energy (about 10 MeV per fission) is not recovered. Overall, about 200 MeV of energy per fission may be recovered for power applications.

Fission Theory

Nuclear fission is a complex process that involves the rearrangement of hundreds of nucleons in a single nucleus to produce two separate nuclei. A complete theoretical understanding of this reaction would require a detailed knowledge of the forces involved in the motion of each of the nucleons through the process. Since such knowledge is still not available, it is necessary to construct simplified models of the actual system to simulate its behavior and gain as accurate a description as possible of the steps in the process. The successes and failures of the models in accounting for the various observations of the fission process can provide new insights into the fundamental physics governing the behavior of real nuclei, particularly at the large nuclear deformations encountered in a nucleus undergoing fission.

The framework for understanding nuclear reactions is analogous to that for chemical reactions and involves the concept of a potential-energy surface on which the reaction occurs. The driving force for physical or chemical reactions is the tendency to lower the potential energy and increase the stability of the system. Thus, for example, a stone at the top of a hill will roll down the hill, converting its potential energy at the top to kinetic energy of motion, and will come to rest at the bottom in a more stable state of lower potential energy. The potential energy is calculated as a function of various parameters of the system being studied. In the case of fission, the potential energy may be calculated as a function of the shape of the system as it proceeds over the barrier to the scission point, and the path of lowest potential energy may be determined.

As has been pointed out, an exact calculation of the nuclear potential energy is not yet possible, and it is to approximate this calculation that various models have been constructed to simulate the real system. Some of the models were developed to address aspects of nuclear structure and spectroscopy as well as features of nuclear reactions, and they also have been employed in attempts to understand the complexity of nuclear fission. The models are based on different assumptions and approximations of the nature of the nuclear forces and the dynamics of the path to scission. No one model can account for all of the extensive phenomenology of fission, but each addresses different aspects of the process and provides a foundation for further development toward a complete theory.

Nuclear Models and Nuclear Fission

The nucleus exhibits some properties that reflect the collective motion of all its constituent nucleons as a unit, as well as other properties that are dependent on the motion and state of the individual nucleons.

The analogy of the nucleus to a drop of an incompressible liquid was first suggested by George Gamow in 1935 and later adapted to a description of nuclear reactions and to fission. Bohr proposed the so-called compound nucleus description of nuclear reactions, in which the excitation energy of the system formed by the absorption of a neutron or photon, for example, is distributed among a large number of degrees of freedom of the system. This excited state persists for a long time relative to the periods of motion of nucleons across the nucleus and then decays by emission of radiation, the evaporation of neutrons or other particles, or by fission. The liquid-drop model of the nucleus accounts quite well for the general collective behavior of nuclei and provides an understanding of the fission process on the basis of the competition between the cohesive nuclear force and the disruptive Coulomb repulsion between protons. It predicts, however, a symmetric division of mass in fission, whereas an asymmetric mass division is observed. Moreover, it does not provide an accurate description of fission barrier systematics or of the ground-state masses of nuclei. The liquid-drop model is particularly useful in describing the behavior of highly excited nuclei, but it does not provide an accurate description for nuclei in their ground or low-lying excited states. Many versions of the liquid-drop model employing improved sets of parameters have been developed. However, investigators have found that mass asymmetry and certain other features in fission cannot be adequately described on the basis of the collective behavior posited by such models alone.

A preference for the formation of unequal masses (i.e., an asymmetric division) was observed early in fission research, and it has remained the most puzzling feature of the process to account for. Investigators have invoked various models other than that of the liquid drop in an attempt to address this question. Dealing with the mutual interaction of all the nucleons in a nucleus has been simplified by treating it as if it were equivalent to the interaction of one particle with an average spherical static potential field that is generated by all the other nucleons. The methods of quantum mechanics provide the solution for the motion of a nucleon in such a potential. A characteristic set of energy levels for neutrons and protons is obtained, and, analogous to the set of levels of the electrons in an atom, the levels group themselves into shells at certain so-called magic numbers of nucleons. (For both neutrons and protons, these numbers are 2, 8, 20, 28, 50, 82, and 126.) Shell closures at these nuclear numbers are marked by especially strong binding, or extra stability. This constitutes the essence of the spherical-shell model (sometimes called the independent-particle, or single-particle, model), as developed by Maria Goeppert Mayer and J. Hans D. Jensen and their colleagues. It accounts well for ground-state masses and spins and for the existence of isomeric nuclear states (excited states having measurable half-lives) that occur when nuclear levels of widely differing spins lie relatively close to each other. The agreement with observations is excellent for spherical nuclei with nucleon numbers near the magic shell numbers. The spherical-shell model, however, does not agree well with the properties of nuclei that have other nucleon numbers—e.g., the nuclei of the lanthanide and actinide elements, with nucleon numbers between the magic numbers.

In the lanthanide and actinide nuclei, the ground state is not spherical but rather deformed into a prolate spheroidal shape—that of a football or watermelon. For such nuclei, the allowed states of motion of a nucleon must be calculated in a potential having a symmetry corresponding to a spheroid rather than a sphere. This was first done by Aage Bohr, Ben R. Mottelson, and Sven G. Nilsson in 1955, and the level structure was calculated as a function of the deformation of the nucleus. A spheroid has three axes of symmetry, and it can rotate in space as a unit about any one of them. The rotation can occur independent of the internal state of excitation of the individual nucleons. Various modes of vibration of the spheroid also may take place. Since this deformed shell model has components of both the independent-particle motion and the collective motion of the nucleus as a whole (i.e., rotations and vibrations), it is sometimes referred to as the unified model.

In Aage Bohr's application of the unified model to the fission process, the sequence of potential-energy surfaces for the excited states of the system are considered to be functions of a deformation parameter (i.e., elongation) characterizing the motion toward fission and evaluated at the saddle point. As the system passes over the saddle point, most of its excitation energy is used up in deforming the nucleus, and the system remains "cold"; i.e., it manifests little excitation, or heat, energy. Thus, only the low-lying excited states are available to the system. The spin and parity of the particular state (or channel) in which the system exists as it passes over the saddle point are then expected to determine the fission properties. In this channel (or transition-state) analysis of fission, a number of characteristics of the process are qualitatively accounted for. Hence, fission thresholds would depend on the spin and parity of the compound nuclear state, the fission fragment angular distribution would be governed by the collective rotational angular momentum of the state, and asymmetry in the mass distribution would result from passage over the barrier in a state of negative parity (which does not possess reflection symmetry). This model gives a good qualitative interpretation of many fission phenomena, but it must assume that at least some of the properties of the transition state at the saddle point are not altered by dynamical considerations in the descent of the system to the scission point. It is the only model that provides a satisfactory interpretation of the angular distributions of fission fragments, and it has attractive features that must be included in any complete theory of fission.

The first application of the spherical-shell model to fission was the recognition that the positions of the peaks in the fission mass distribution correlated fairly well with the magic numbers and suggested a qualitative interpretation of the asymmetric mass division. Thus, a preference for the formation of nuclei with neutron numbers close to 82 would favour the formation of nuclides near the peak in the heavy group and would thus determine the mass split for the fissioning system . Some extra stability for nuclear configurations of 50 protons would also be expected, but this is not particularly evident. In fact, the so-called doubly magic nucleus tin-132, with 50 protons and 82 neutrons, has a rather low yield in low-energy fission.

A more quantitative application of the spherical-shell model to fission was undertaken by Peter Fong in the United States in 1956. He related the probability of formation of a given pair of fragments to the available density of states for that pair of fragments at the scission point in a statistical-model approach. A model of this sort predicts that the system, in its random motions, will experience all possible configurations and so will have a greater probability of being in the region where the greatest number of such configurations (or states) is concentrated. The model assumes that the potential energy at the saddle point is essentially all converted to excitation energy and that a statistical equilibrium among all possible states is established at the scission point. The extra binding energy for closed-shell nuclei leads to a higher density of states at a given excitation energy than is present for other nuclei and, hence, leads to a higher probability of formation. An asymmetric mass distribution in good agreement with that observed for the neutron-induced fission of uranium-235 is obtained. Moreover, the changes in the mass distribution with an increased excitation energy of fission (e.g., an increase in the probability of symmetric fission relative to asymmetric fission) are accounted for by the decrease in importance of the shell effects as the excitation energy increases. Other features of the fission process also are qualitatively explained; however, extensive changes in the parameters of the model are required to obtain agreement with experiments for other fissionable nuclides. Then, too, there are fundamental problems concerning the validity of some of the basic assumptions of the model.

The fundamental question as to the validity of models that evaluate the properties of the system at the scission point (the so-called scission-point models of fission) is whether the system remains long enough at this point on the steep decline of the potential-energy surface for a quasi-equilibrium condition to be established. There is some evidence that such a condition may indeed prevail, but it is not clearly established. Nonetheless, such models have proved quite useful in interpreting observations of mass, charge, and kinetic energy distributions, as well as of neutron emission dependence on fragment mass. It seems very likely that the fragment shell structure plays a significant role in determining the course of the fission process.

Although the single-particle models provide a good description of various aspects of nuclear structure, they are not successful in accounting for the energy of deformation of nuclei (i.e., surface energy), particularly at the large deformations encountered in the fission process. A major breakthrough occurred when a hybrid model incorporating shell effects as a correction to the potential energy of the liquid-drop model was proposed by the Russian physicist V.M. Strutinskii in 1967. This approach retains the dominant collective surface and Coulomb effects while adding shell and pairing corrections that depend on deformation. Shell corrections of several million electron volts are calculated, and these can have a significant effect on a liquid-drop barrier of about 5 MeV. The nucleon numbers at which the shells appear depend on the deformation and may differ from the spherical model magic numbers. In the vicinity of the fission barrier, the shells introduce structure in the liquid-drop potential-energy curve, as illustrated in figure. The relative heights and widths of the two peaks vary with the mass and charge of the fissioning system.

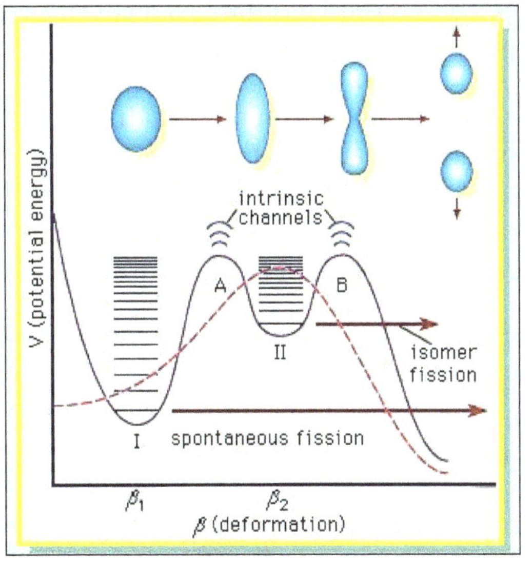

In figure, schematic illustrations of single-humped and double-humped fission barriers. The former are represented by the dashed line and the latter by the continuous line. Intrinsic excitations in the first and second wells at deformations β_1 and β_2 are designated class I and class II states, respectively. Intrinsic channels at the two barriers also are illustrated. The transition in the shape of the nucleus as a function of deformation is schematically represented in the upper part of the figure. Spontaneous fission of the ground state and isomeric state occurs from the lowest energy class I and class II states, respectively.

The double-humped barrier provides a satisfactory explanation for a number of puzzling observations in fission. The existence of short-lived, spontaneous fission isomers, for example, is understood as the consequence of the population of states in the second well (class II). These isomers have a much smaller barrier to penetrate and so exhibit a much shorter spontaneous fission half-life. The change in shape associated with these states, as compared to class I states, also hinders a rapid return to the ground state by gamma emission. (Class II states are also called shape isomers.) The systematics of neutron-induced fission cross sections and structure in some fission-fragment angular distributions also find an interpretation in the implications of the double-humped barrier.

The Strutinskii procedure provided a strong stimulus for calculations of the potential-energy surfaces appropriate to fissioning systems, since it provided a consistent and useful prescription for treating both the macroscopic (liquid-drop) and microscopic (single-particle) effects in deformed nuclei. Many calculations of the potential-energy surface employing different model potentials and parameters have been carried out as functions of the shapes of the system. The work of the American nuclear physicists W.J. Swiatecki, James R. Nix, and their collaborators has been particularly noteworthy in such studies, which also include some attempts to treat the dynamical evolution of the fission process.

Calculations for the actinide elements indicate that, at deformations corresponding to the second barrier, the potential energy for asymmetric mass splits is lower than that for symmetric ones; hence, the former are favoured at that stage of the process. For larger deformations, however, a single potential does not represent the incipient formation of two fragments very well. In fact, a discontinuity occurs at the scission point, and the results of the calculation depend on whether the scission configuration is treated as one nucleus or as two separate nuclei.

A two-centre potential may also be used to represent the nature of the forces at work in a fissioning nucleus. In such a model, the potential energy surfaces are represented by two overlapping spheres or spheroids. It is equivalent to a one-centre potential when there is a complete overlap at small deformations, and it has the correct asymptotic behavior as the nascent fragments separate. This approach indicates a preformation of the final shell structure of the fragments early in the process.

Although the validity of the assumptions inherent in scission-point models may be in question, the results obtained with them are in excellent agreement with observation. Representative of such a model is the Argonne Scission-Point model, which uses a macroscopic-microscopic calculation with deformed fragment shell and pairing corrections to determine the potential energy of a system of two nearly touching spheroids and which includes their interaction in terms of a neck connecting them. Models of this kind provide a simple approach to a highly quantitative and detailed study of the dependence of the probability of formation of a given fragment pair on the neutron and proton number and on the deformation in each fragment. They account very well for the mass, charge, and kinetic-energy distributions and the neutron-emission dependence on mass number for a broad range of fissioning nuclei. The scission-point models, however, do not address questions of fission probability or the angular distributions of the fragments. As the fission-excitation energy increases, the shell correction diminishes and the macroscopic (liquid-drop) behavior dominates.

Nuclides in the region of fermium-264 have been observed to undergo symmetric fission with unusually high fragment kinetic energies. This appears to be the consequence of the stability for the magic number configurations of 50 protons and 82 neutrons. The formation of two doubly magic fragments of tin-132 is strongly favoured energetically, whereas the formation of only one such fragment in the low-energy fission of uranium or plutonium isotopes is not. The fragments of tin-132 are spherical rather than deformed, and a more compact configuration at the scission point (with the charge centres closer together) leads to higher fragment kinetic energies.

It is evident that shell effects, both in the fissioning system at the saddle point and in the deformed fragments near the scission point, are important in interpreting many of the features of the fission process. The stage of the process at which the various fragment distributions are determined is, however, not clearly established. All the components of a reasonable understanding of fission seem to be at hand, but they have yet to be synthesized into a complete, dynamic theory.

Considerations of the dynamics of the descent of the system on the potential-energy surface from the saddle point to the scission point involve two extreme points of view. An "adiabatic" approximation may be valid if the collective motion of the system is considered to be so slow—or the coupling between the collective and internal single-particle degrees of freedom (i.e., between macroscopic and microscopic behavior) so weak—that the fast single-particle motions can readily adjust to the changes in shape of the fissioning nucleus as it progresses toward scission. In this case, the changes in the system take place without the gain or loss of heat energy. The decrease in potential energy between the saddle and scission points will then appear primarily in the collective degrees of freedom at scission and be associated with the kinetic energy of the relative motion of the nascent fragments (referred to as pre-scission kinetic energy). On the other hand, if the collective motion toward scission is relatively fast or the coupling-to-particle motion stronger, collective energy can be transformed into internal excitation (heat) energy of the nucleons. (This is analogous to heating in the motion of a viscous fluid.) In such a "non-adiabatic" process the mixing among the single-particle degrees of freedom may be sufficiently complete that a statistical model may be applicable at the scission point. Either extreme represents an approximation of complex behavior, and some experimental evidence in support of either interpretation may be advanced. As in most such instances in nature, the truth probably lies somewhere between the extremes, with both playing some role in the fission process.

Fission Chain Reactions and their Control

The emission of several neutrons in the fission process leads to the possibility of a chain reaction if at least one of the fission neutrons induces fission in another fissile nucleus, which in turn fissions and emits neutrons to continue the chain. If more than one neutron is effective in inducing fission in other nuclei, the chain multiplies more rapidly. The condition for a chain reaction is usually expressed in terms of a multiplication factor, k, which is defined as the ratio of the number of fissions produced in one step (or neutron generation) in the chain to the number of fissions in the preceding generation. If k is less than unity, a chain reaction cannot be sustained. If $k = 1$, a steady-state chain reaction can be maintained; and if k is greater than 1, the number of fissions increases at each step, resulting in a divergent chain reaction. The term critical assembly is applied to a configuration of fissionable material for which $k = 1$; if $k > 1$, the assembly is said to be supercritical. A critical assembly might consist of the fissile material in the form of a metal or oxide, a moderator to slow the fission neutrons, and a reflector to scatter neutrons that would otherwise be lost back into the assembly core.

In a fission bomb it is desirable to have k as large as possible and the time between steps in the chain as short as possible so that many fissions occur and a large amount of energy is generated within a brief period ($\sim 10^{-7}$ second) to produce a devastating explosion. If one kilogram of uranium-235 were to fission, the energy released would be equivalent to the explosion of 20,000 tons of the chemical explosive trinitrotoluene (TNT). In a controlled nuclear reactor, k is kept equal to unity for steady-state operation. A practical reactor, however, must be designed with k somewhat greater than unity. This permits power levels to be increased if desired, as well as allowing for the following: the gradual loss of fuel by the fission process; the buildup of "poisons" among the fission products being formed that absorb neutrons and lower the k value; and the use of some of the neutrons produced for research studies or the preparation of radioactive species for various applications. The value of k is controlled during the operation of a reactor by the positioning of movable rods made of a material that readily absorbs neutrons (i.e., one with a high neutron-capture cross section), such as boron, cadmium, or hafnium. The delayed-neutron emitters among the fission products increase the time between successive neutron generations in the chain reaction and make the control of the reaction easier to accomplish by the mechanical movement of the control rods.

Fission reactors can be classified by the energy of the neutrons that propagate the chain reaction. The most common type, called a thermal reactor, operates with thermal neutrons (those having the same energy distribution as gas molecules at ordinary room temperatures). In such a reactor, the fission neutrons produced (with an average kinetic energy of more than 1 MeV) must be slowed down to thermal energy by scattering from a moderator, usually consisting of ordinary water, heavy water (D_2O), or graphite. In another type, termed an intermediate reactor, the chain reaction is maintained by neutrons of intermediate energy, and a beryllium moderator may be used. In a fast reactor, fast fission neutrons maintain the chain reaction, and no moderator is needed. All of the reactor types require a coolant to remove the heat generated; water, a gas, or a liquid metal may be used for this purpose, depending on the design needs.

Uses of Fission Reactors and Fission Products

A nuclear reactor is essentially a furnace used to produce steam or hot gases that can provide heat directly or drive turbines to generate electricity. Nuclear reactors are employed for commercial electric-power generation throughout much of the world and as a power source for propelling submarines and certain kinds of surface vessels. Another important use for reactors is the utilization of their high neutron fluxes for studying the structure and properties of materials and for producing a broad range of radionuclides, which, along with a number of fission products, have found many different applications. Heat generated by radioactive decay can be converted into electricity through the thermoelectric effect in semiconductor materials and thereby produce what is termed an atomic battery. When powered by either a long-lived beta-emitting fission product (e.g., strontium-90 or promethium-147) or one that emits alpha particles (plutonium-238 or curium-244), these batteries are a particularly useful source of energy for cardiac pacemakers and for instruments employed in remote unmanned facilities, such as those in outer space, the polar regions of the Earth, or the open seas.

Nuclear Fusion

Nuclear fusion is the process by which nuclear reactions between light elements form heavier elements (up to iron). In cases where the interacting nuclei belong to elements with low atomic numbers (e.g., hydrogen [atomic number 1] or its isotopes deuterium and tritium), substantial amounts of energy are released. The vast energy potential of nuclear fusion was first exploited in thermonuclear weapons, or hydrogen bombs, which were developed in the decade immediately following World War II. Meanwhile, the potential peaceful applications of nuclear fusion, especially in view of the essentially limitless supply of fusion fuel on Earth, have encouraged an immense effort to harness this process for the production of power.

The Fusion Reaction

Fusion reactions constitute the fundamental energy source of stars, including the Sun. The evolution of stars can be viewed as a passage through various stages as thermonuclear reactions and nucleosynthesis cause compositional changes over long time spans. Hydrogen (H) "burning" initiates the fusion energy source of stars and leads to the formation of helium (He). Generation of fusion energy for practical use also relies on fusion reactions between the lightest elements that burn to form helium. In fact, the heavy isotopes of hydrogen—deuterium (D) and tritium (T)—react more efficiently with each other, and, when they do undergo fusion, they yield more energy per reaction than do two hydrogen nuclei. (The hydrogen nucleus consists of a single proton. The deuterium nucleus has one proton and one neutron, while tritium has one proton and two neutrons).

Fusion reactions between light elements, like fission reactions that split heavy elements, release energy because of a key feature of nuclear matter called the binding energy, which can be released through fusion or fission. The binding energy of the nucleus is a measure of the efficiency with which its constituent nucleons are bound together. Take, for example, an element with Z protons and N neutrons in its nucleus. The element's atomic weight A is Z + N, and its atomic number is Z. The binding energy B is the energy associated with the mass difference between the Z protons and N neutrons considered separately and the nucleons bound together (Z + N) in a nucleus of mass M. The formula is:

$$B = (Zm_p + Nm_n - M)c^2,$$

Where, m_p and m_n are the proton and neutron masses and c is the speed of light. It has been determined experimentally that the binding energy per nucleon is a maximum of about 1.4 10−12 joule at an atomic mass number of approximately 60—that is, approximately the atomic mass number of iron. Accordingly, the fusion of elements lighter than iron or the splitting of heavier ones generally leads to a net release of energy.

Two Types of Fusion Reactions

Fusion reactions are of two basic types: (1) those that preserve the number of protons and neutrons and (2) those that involve a conversion between protons and neutrons. Reactions of the first type are most important for practical fusion energy production, whereas those of the second type are crucial to the initiation of star burning. An arbitrary element is indicated by the notation A_ZX, where Z is the charge of the nucleus and A is the atomic weight. An important fusion reaction for practical energy generation is that between deuterium and tritium (the D-T fusion reaction). It produces helium (He) and a neutron (n) and is written:

$$D + T \rightarrow He + n.$$

To the left of the arrow (before the reaction) there are two protons and three neutrons. The same is true on the right.

The other reaction, that which initiates star burning, involves the fusion of two hydrogen nuclei to form deuterium (the H-H fusion reaction):

$$H + H \rightarrow D + \beta^+ + \nu,$$

where β^+ represents a positron and ν stands for a neutrino. Before the reaction there are two hydrogen nuclei (that is, two protons). Afterward there are one proton and one neutron (bound together as the nucleus of deuterium) plus a positron and a neutrino (produced as a consequence of the conversion of one proton to a neutron).

Both of these fusion reactions are exoergic and so yield energy. The German-born physicist Hans Bethe proposed in the 1930s that the H-H fusion reaction could occur with a net release of energy and provide, along with subsequent reactions, the fundamental energy source sustaining the stars. However, practical energy generation requires the D-T reaction for two reasons: first, the rate of reactions between deuterium and tritium is much higher than that between protons; second, the net energy release from the D-T reaction is 40 times greater than that from the H-H reaction.

Energy Released in Fusion Reactions

Energy is released in a nuclear reaction if the total mass of the resultant particles is less than the mass of the initial reactants. To illustrate, suppose two nuclei, labeled X and a, react to form two other nuclei, Y and b, denoted:

$$X + a \rightarrow Y + b.$$

The particles a and b are often nucleons, either protons or neutrons, but in general can be any nuclei. Assuming that none of the particles is internally excited (i.e., each is in its ground state), the energy quantity called the Q-value for this reaction is defined as:

$$Q = (m_x + m_a - m_b - m_y)c^2,$$

where, the m-letters refer to the mass of each particle and c is the speed of light. When the energy value Q is positive, the reaction is exoergic; when Q is negative, the reaction is endoergic (i.e., absorbs energy). When both the total proton number and the total neutron number are preserved before and after the reaction (as in D-T reactions), then the Q-value can be expressed in terms of the binding energy B of each particle as:

$$Q = B_y + B_b - B_x - B_a.$$

The D-T fusion reaction has a positive Q-value of 2.8×10^{-12} joule. The H-H fusion reaction is also exoergic, with a Q-value of 6.7×10^{-14} joule. To develop a sense for these figures, one might consider that one metric ton (1,000 kg, or almost 2,205 pounds) of deuterium would contain roughly 3×10^{32} atoms. If one ton of deuterium were to be consumed through the fusion reaction with tritium, the energy released would be 8.4×10^{20} joules. This can be compared with the energy content of one ton of coal—namely, 2.9×10^{10} joules. In other words, one ton of deuterium has the energy equivalent of approximately 29 billion tons of coal.

Rate and Yield of Fusion Reactions

The energy yield of a reaction between nuclei and the rate of such reactions are both important. These quantities have a profound influence in scientific areas such as nuclear astrophysics and the potential for nuclear production of electrical energy.

When a particle of one type passes through a collection of particles of the same or different type, there is a measurable chance that the particles will interact. The particles may interact in many ways, such as simply scattering, which means that they change direction and exchange energy, or they may undergo a nuclear fusion reaction. The measure of the likelihood that particles will interact is called the cross section, and the magnitude of the cross section depends on the type of interaction and the state and energy of the particles. The product of the cross section and the atomic density of the target particle is called the macroscopic cross section. The inverse of the macroscopic cross section is particularly noteworthy as it gives the mean distance an incident particle will travel before interacting with a target particle; this inverse measure is called the mean free path. Cross sections are measured by producing a beam of one particle at a given energy, allowing the beam to interact with a (usually thin) target made of the same or a different material, and measuring deflections or reaction products. In this way it is possible to determine the relative likelihood of one type of fusion reaction versus another, as well as the optimal conditions for a particular reaction.

The cross sections of fusion reactions can be measured experimentally or calculated theoretically, and they have been determined for many reactions over a wide range of particle energies. They are well known for practical fusion energy applications and are reasonably well known, though with gaps, for stellar evolution. Fusion reactions between nuclei, each with a positive charge of one or more, are the most important for both practical applications and the nucleosynthesis of the light elements in the burning stages of stars. Yet, it is well known that two positively charged nuclei repel each other electrostatically—i.e., they experience a repulsive force inversely proportional to the square of the distance separating them. This repulsion is called the Coulomb barrier It is highly unlikely that two positive nuclei will approach each other closely enough to undergo a fusion reaction unless they have sufficient energy to overcome the Coulomb barrier. As a result, the cross section for fusion reactions between charged particles is very small unless the energy of the particles is high, at least 10^4 electron volts (1 eV ≅ 1.602 × 10^{-19} joule) and often more than 10^5 or 10^6 eV. This explains why the centre of a star must be hot for the fuel to burn and why fuel for practical fusion energy systems must be heated to at least 50,000,000 kelvins (K; 90,000,000 °F). Only then will a reasonable fusion reaction rate and power output be achieved.

The phenomenon of the Coulomb barrier also explains a fundamental difference between energy generation by nuclear fusion and nuclear fission. While fission of heavy elements can be induced by either protons or neutrons, generation of fission energy for practical applications is dependent on neutrons to induce fission reactions in uranium or plutonium. Having no electric charge, the neutron is free to enter the nucleus even if its energy corresponds to room temperature. Fusion energy, relying as it does on the fusion reaction between light nuclei, occurs only when the particles are sufficiently energetic to overcome the Coulomb repulsive force. This requires the production and heating of the gaseous reactants to the high temperature state known as the plasma state.

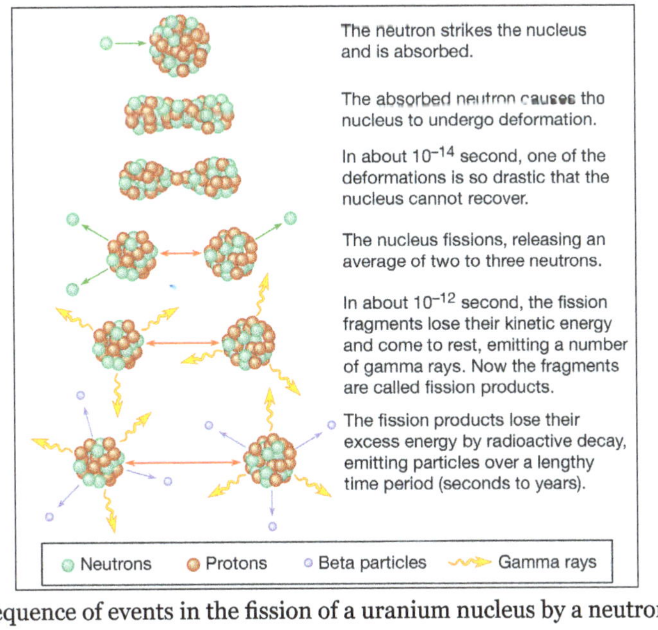

Sequence of events in the fission of a uranium nucleus by a neutron.

The Plasma State

Typically, a plasma is a gas that has had some substantial portion of its constituent atoms or molecules ionized by the dissociation of one or more of their electrons. These free electrons enable plasmas to conduct electric charges, and a plasma is the only state of matter in which thermonuclear reactions can occur in a self-sustaining manner. Astrophysics and magnetic fusion research, among other fields, require extensive knowledge of how gases behave in the plasma state. The stars, the solar wind, and much of interstellar space are examples where the matter present is in the plasma state. Very high-temperature plasmas are fully ionized gases, which means that the ratio of neutral gas atoms to charged particles is small. For example, the ionization energy of hydrogen is 13.6 eV, while the average energy of a hydrogen ion in a plasma at 50,000,000 K is 6,462 eV. Thus, essentially all of the hydrogen in this plasma would be ionized.

A reaction-rate parameter more appropriate to the plasma state is obtained by accounting for the fact that the particles in a plasma, as in any gas, have a distribution of energies. That is to say, not all particles have the same energy. In simple plasmas this energy distribution is given by the Maxwell-Boltzmann distribution law, and the temperature of the gas or plasma is, within a proportionality constant, two-thirds of the average particle energy; i.e., the relationship between the average energy $\bar{E}$ and temperature T is $\bar{E} = 3kT/2$, where k is the Boltzmann constant, 8.62×10^{-5} eV per kelvin. The intensity of nuclear fusion reactions in a plasma is derived by averaging the product of the particles' speed and their cross sections over a distribution of speeds corresponding to a Maxwell-Boltzmann distribution. The cross section for the reaction depends on the energy or speed of the particles. The averaging process yields a function for a given reaction that depends only on the temperature and can be denoted f(T). The rate of energy released (i.e., the power released) in a reaction between two species, a and b, is:

$$P_{ab} = n_a n_b f_{ab}(T) U_{ab},$$

where, n_a and n_b are the density of species a and b in the plasma, respectively, and U_{ab} is the energy released each time a and b undergo a fusion reaction. The parameter Pab properly takes into account both the rate of a given reaction and the energy yield per reaction.

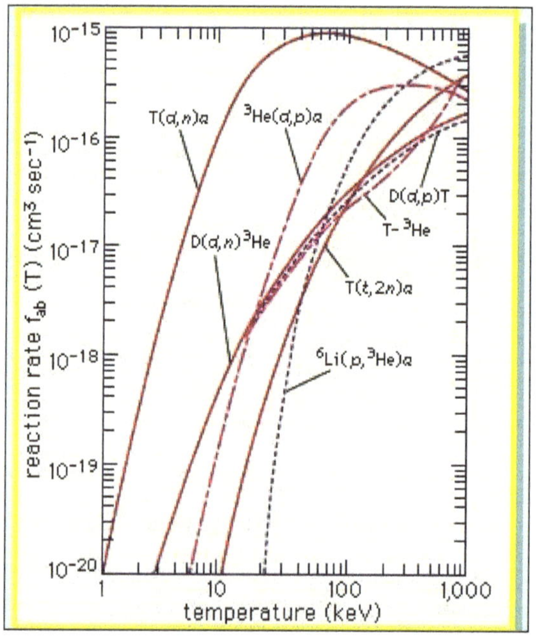

The reaction rate as a function of plasma temperature, expressed in kiloelectron volts (keV; 1 keV is equivalent to a temperature of 11,000,000 K). The rate of reaction between deuterium and tritium is seen to be higher than all others and is very substantial, even at temperatures in the 5-to-10-keV range.

Fusion Reactions in Stars

Fusion reactions are the primary energy source of stars and the mechanism for the nucleosynthesis of the light elements. In the late 1930s Hans Bethe first recognized that the fusion of hydrogen nuclei to form deuterium is exoergic (i.e., there is a net release of energy) and, together with subsequent nuclear reactions, leads to the synthesis of helium. The formation of helium is the main source of energy emitted by normal stars, such as the Sun, where the burning-core plasma has a temperature of less than 15,000,000 K. However, because the gas from which a star is formed often contains some heavier elements, notably carbon (C) and nitrogen (N), it is important to include nuclear reactions between protons and these nuclei. The reaction chain between protons that ultimately leads to helium is the proton-proton cycle. When protons also induce the burning of carbon and nitrogen, the CN cycle must be considered; and, when oxygen (O) is included, still another alternative scheme, the CNO bi-cycle, must be accounted for.

The proton-proton nuclear fusion cycle in a star containing only hydrogen begins with the reaction,

$$H + H \rightarrow D + \beta^{+} + \nu;\ Q = 1.44\ \text{MeV},$$

where the Q-value assumes annihilation of the positron by an electron. The deuterium could react with other deuterium nuclei, but, because there is so much hydrogen, the D/H ratio is held to very low values, typically 10–18. Thus, the next step is,

$$H + D \rightarrow {}^{3}He + \gamma;\ Q = 5.49\ \text{MeV},$$

where γ indicates that gamma rays carry off some of the energy yield. The burning of the helium-3 isotope then gives rise to ordinary helium and hydrogen via the last step in the chain:

$${}^{3}He + {}^{3}He \rightarrow {}^{4}He + 2(H);\ Q = 12.86\ \text{MeV}.$$

At equilibrium, helium-3 burns predominantly by reactions with itself because its reaction rate with hydrogen is small, while burning with deuterium is negligible due to the very low deuterium concentration. Once helium-4 builds up, reactions with helium-3 can lead to the production of still-heavier elements, including beryllium-7, beryllium-8, lithium-7, and boron-8, if the temperature is greater than about 10,000,000 K.

The stages of stellar evolution are the result of compositional changes over very long periods. The size of a star, on the other hand, is determined by a balance between the pressure exerted by the hot plasma and the gravitational force of the star's mass. The energy of the burning core is transported toward the surface of the star, where it is radiated at an effective temperature. The effective temperature of the Sun's surface is about 6,000 K, and significant amounts of radiation in the visible and infrared wavelength ranges are emitted.

Fusion Reactions for Controlled Power Generation

Reactions between deuterium and tritium are the most important fusion reactions for controlled power generation because the cross sections for their occurrence are high, the practical plasma temperatures required for net energy release are moderate, and the energy yield of the reactions are high—17.58 MeV for the basic D-T fusion reaction.

It should be noted that any plasma containing deuterium automatically produces some tritium and helium-3 from reactions of deuterium with other deuterium ions. Other fusion reactions involving elements with an atomic number above 2 can be used, but only with much greater difficulty. This is because the Coulomb barrier increases with increasing charge of the nuclei, leading to the requirement that the plasma temperature exceed 1,000,000,000 K if a significant rate is to be achieved. Some of the more interesting reactions are:

1. $H + {}^{11}B \rightarrow 3({}^{4}He)$; Q = 8.68 MeV;
2. $H + {}^{6}Li \rightarrow {}^{3}He + {}^{4}He$; Q = 4.023 MeV;
3. ${}^{3}He + {}^{6}Li \rightarrow H + 2({}^{4}He)$; Q = 16.88 MeV; and
4. ${}^{3}He + {}^{6}Li \rightarrow D + {}^{7}Be$; Q = 0.113 MeV.

Reaction 2nd converts lithium-6 to helium-3 and ordinary helium. Interestingly, if reaction 2nd is followed by reaction 3rd, then a proton will again be produced and be available to induce reaction 2nd, thereby propagating the process. Unfortunately, it appears that reaction 4th is 10 times more likely to occur than reaction 3rd.

Methods of Achieving Fusion Energy

Practical efforts to harness fusion energy involve two basic approaches to containing a high-temperature plasma of elements that undergo nuclear fusion reactions: magnetic confinement and inertial confinement. A much less likely but nevertheless interesting approach is based on fusion catalyzed by muons.

Magnetic Confinement

In magnetic confinement the particles and energy of a hot plasma are held in place using magnetic fields. A charged particle in a magnetic field experiences a Lorentz force that is proportional to the product of the particle's velocity and the magnetic field. This force causes electrons and ions to spiral about the direction of the magnetic line of force, thereby confining the particles. When the topology of the magnetic field yields an effective magnetic well and the pressure balance between the plasma and the field is stable, the plasma can be confined away from material boundaries. Heat and particles are transported both along and across the field, but energy losses can be prevented in two ways. The first is to increase the strength of the magnetic field at two locations along the field line. Charged particles contained between these points can be made to reflect back and forth, an effect called magnetic mirroring. In a basically straight system with a region of intensified magnetic field at each end, particles can still escape through the ends due to scattering between particles as they approach the mirroring points. Such end losses can be avoided altogether by creating a magnetic field in the topology of a torus (i.e., configuration of a doughnut or inner tube).

External magnets can be arranged to create a magnetic field topology for stable plasma confinement, or they can be used in conjunction with magnetic fields generated by currents induced to flow in the plasma itself. The late 1960s witnessed a major advance by the Soviet Union in harnessing fusion reactions for practical energy production. Soviet scientists achieved a high plasma temperature (about 3,000,000 K), along with other physical parameters, in a machine referred to as a tokamak.

A tokamak is a toroidal magnetic confinement system in which the plasma is kept stable both by an externally generated, doughnut-shaped magnetic field and by electric currents flowing within the plasma. Since the late 1960s the tokamak has been the major focus of magnetic fusion research worldwide, though other approaches such as the stellarator, the compact torus, and the reversed field pinch (RFP) have also been pursued. In these approaches, the magnetic field lines follow a helical, or screwlike, path as the lines of magnetic force proceed around the torus. In the tokamak the pitch of the helix is weak, so the field lines wind loosely around the poloidal direction (through the central hole) of the torus. In contrast, RFP field lines wind much tighter, wrapping many times in the poloidal direction before completing one loop in the toroidal direction (around the central hole).

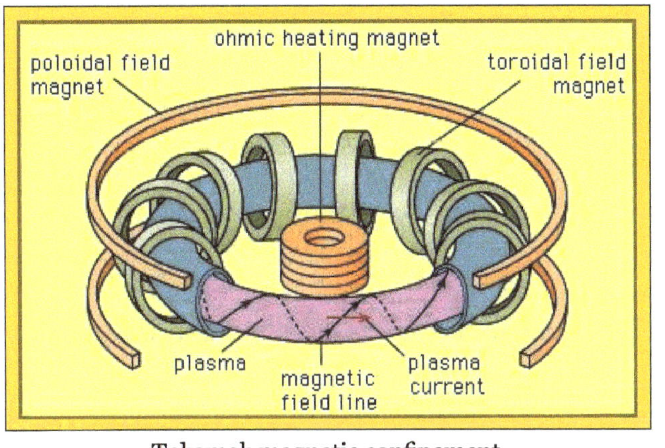

Tokamak magnetic confinement.

Magnetically confined plasma must be heated to temperatures at which nuclear fusion is vigorous, typically greater than 75,000,000 K (equivalent to an energy of 4,400 eV). This can be achieved by coupling radio-frequency waves or microwaves to the plasma particles, by injecting energetic beams of neutral atoms that become ionized and heat the plasma, by magnetically compressing the plasma, or by the ohmic heating (also known as Joule heating) that occurs when an electric current passes through the plasma.

Employing the tokamak concept, scientists and engineers in the United States, Europe, and Japan began in the mid-1980s to use large experimental tokamak devices to attain conditions of temperature, density, and energy confinement that now match those necessary for practical fusion power generation. The machines employed to achieve these results include the Joint European Torus (JET) of the European Union, the Japanese Tokamak-60 (JT-60), and, until 1997, the Tokamak Fusion Test Reactor (TFTR) in the United States. Indeed, in both the TFTR and the JET devices, experiments using deuterium and tritium produced more than 10 megawatts of fusion power and essentially energy breakeven conditions in the plasma itself. Plasma conditions approaching those achieved in tokamaks were also achieved in large stellarator machines in Germany and Japan during the 1990s.

Inertial Confinement Fusion (ICF)

In this approach, a fuel mass is compressed rapidly to densities 1,000 to 10,000 times greater than normal by generating a pressure as high as 10^{17} pascals (10^{12} atmospheres) for periods as short as a nanosecond (10^{-9} second). Near the end of this time period, the implosion speed exceeds about 3×10^5 metres per second. At maximum compression of the fuel, which is now in a cool plasma state, the energy in converging shock waves is sufficient to heat the very centre of the fuel to temperatures high enough to induce fusion reactions (greater than an equivalent energy of about 4,400 eV). If the mass of this highly compressed fuel material is large enough, energy will be generated through fusion reactions before this hot plasma ball disassembles. Under proper conditions, much more energy can be released than is required to compress and shock heat the fuel to thermonuclear burning conditions.

The physical processes in ICF bear a relationship to those in thermonuclear weapons and in star formation—namely, collapse, compression heating, and the onset of nuclear fusion. The situation in star formation differs in one respect: gravity is the cause of the collapse, and a collapsed star begins to expand again due to heat from exoergic nuclear fusion reactions. The expansion is ultimately arrested by the gravitational force associated with the enormous mass of the star, at which point a state of equilibrium in both size and temperature is achieved. In contrast, the fuel in a thermonuclear weapon or ICF completely disassembles. In the ideal ICF case, however, this does not occur until about 30 percent of the fusion fuel has burned.

Over the decades, very significant progress has been made in developing the technology and systems for high-energy, short-time-pulse drivers that are necessary to implode the fusion fuel. The most common driver is a high-power laser, though particle accelerators capable of producing beams of high-energy ions are also used. Lasers that produce more than 100,000 joules in pulses of about one nanosecond are now used in experiments, and the power available in short bursts exceeds 10^{14} watts.

Two lasers capable of delivering up to 5,000,000 joules in equally short bursts, generating a power level on the fusion targets in excess of 5×10^{14} watts, are operational. One facility is the Laser MegaJoule in Bordeaux, France. The other is the National Ignition Facility at the Lawrence Livermore National Laboratory in Livermore, Calif.

Muon-catalyzed Fusion

The need in traditional schemes of nuclear fusion to confine very high-temperature plasmas has led some researchers to explore alternatives that would permit fusion reactants to approach each other more closely at much lower temperatures. One method involves substituting muons (μ) for the electrons that ordinarily surround the nucleus of a fuel atom. Muons are negatively charged subatomic particles similar to electrons, except that their mass is a little more than 200 times the electron mass and they are unstable, having a half-life of about 2.2×10^{-6} second. In fact, fusion has been observed in liquid and gas mixtures of deuterium and tritium at cryogenic temperatures when muons were injected into the mixture.

Muon-catalyzed fusion is the name given to the process of achieving fusion reactions by causing a deuteron (deuterium nucleus, D^+), a triton (tritium nucleus, T^+), and a muon to form what is called a muonic molecule. Once a muonic molecule is formed, the rate of fusion reactions is approximately 3×10^{-8} second. However, the formation of a muonic molecule is complex, involving a series of atomic, molecular, and nuclear processes.

In schematic terms, when a muon enters a mixture of deuterium and tritium, the muon is first captured by one of the two hydrogen isotopes in the mixture, forming either atomic D^+-μ or T^+-μ, with the atom now in an excited state. The excited atom relaxes to the ground state through a cascade collision process, in which the muon may be transferred from a deuteron to a triton or vice versa. More important, it is also possible that a muonic molecule (D^+-μ-T^+) will be formed. Although a much rarer reaction, once a muonic molecule does form, fusion takes place almost immediately, releasing the muon in the mixture to be captured again by a deuterium or tritium nucleus and allowing the process to continue. In this sense the muon acts as a catalyst for fusion reactions within the mixture. The key to practical energy production is to generate enough fusion reactions before the muon decays.

The complexities of muon-catalyzed fusion are many and include generating the muons (at an energy expenditure of about five billion electron volts per muon) and immediately injecting them into the deuterium-tritium mixture. In order to produce more energy than what is required to initiate the process, about 300 D-T fusion reactions must take place within the half-life of a muon.

Cold Fusion and Bubble Fusion

Two disputed fusion experiments merit mention. In 1989 two chemists, Martin Fleischmann of the University of Utah and Stanley Pons of the University of Southampton in England, announced that they had produced fusion reactions at essentially room temperature. Their system consisted of electrolytic cells containing heavy water (deuterium oxide, D_2O) and palladium rods that absorbed the deuterium from the heavy water. Efforts to give a theoretical explanation of the results failed, as did worldwide efforts to reproduce the claimed cold fusion.

In 2002 Rusi Taleyarkhan and colleagues at Purdue University in Lafayette, Ind., claimed to have observed a statistically significant increase in nuclear emissions of products of fusion reactions (neutrons and tritium) during acoustic cavitation experiments with chilled deuterated (bombarded with deuterium) acetone. Their experimental setup was based on the known phenomenon of sonoluminescence. In sonoluminescence a gas bubble is imploded with high-pressure sound waves. At the end of the implosion process, and for a short time afterward, conditions of high density and temperature are achieved that lead to light emission. By starting with larger, millimetre-sized cavitations (bubbles) that had been deuterated in the acetone liquid, the researchers claimed to have produced densities and temperatures sufficient to induce fusion reactions just before the bubbles broke up. As with cold fusion, most attempts to replicate their results have failed.

Conditions for Practical Fusion Yield

Two conditions must be met to achieve practical energy yields from fusion. First, the plasma temperature must be high enough that fusion reactions occur at a sufficient rate. Second, the plasma must be confined so that the energy released by fusion reactions, when deposited in the plasma, maintains its temperature against loss of energy by such phenomena as conduction, convection, and radiation. When these conditions are achieved, the plasma is said to be ignited. In the case of stars, or some approaches to fusion by magnetic confinement, a steady state can be achieved, and no energy beyond what is supplied from fusion reactions is needed to sustain the system. In other cases, such as the ICF approach, there is a large temperature excursion once fuel ignition is achieved. The energy yield can far exceed the energy required to attain plasma ignition conditions, but this energy is released in a burst, and the process has to be repeated roughly once every second for practical power to be produced.

The conditions for plasma ignition are readily derived. When fusion reactions occur in a plasma, the power released is proportional to the square of plasma ion density, n^2. The plasma loses energy when electrons scatter from positively charged ions, accelerating and radiating in the process. Such radiation is called bremsstrahlung and is proportional to $n^2T^{1/2}$, where T is the plasma temperature. Other mechanisms by which heat can escape the plasma lead to a characteristic energy-loss time denoted by τ. The energy content of the plasma at temperature T is 3nkT, where k is the Boltzmann constant. The rate of energy loss by mechanisms other than bremsstrahlung is thus simply $3nkT/\tau$. The energy balance of the plasma is the balance between the fusion energy heating the plasma and the energy-loss rate, which is the sum of $3nkT/\tau$ and the bremsstrahlung. The condition satisfying this balance is called the ignition condition. An equation relates the product of density and energy confinement time, denoted $n\tau$, to a function that depends only on the plasma temperature and the type of fusion reaction. For example, when the plasma is composed of deuterium and tritium, the smallest value of $n\tau$ required to achieve ignition is about 2×10^{20} particles per cubic metre times seconds, and the required temperature corresponds to an energy of about 25,000 eV. If the only energy losses are due to bremsstrahlung escaping from the plasma (meaning τ is infinite), the ignition temperature decreases to an energy level of 4,400 eV. Hence, the keys to generating usable amounts of fusion energy are to attain a sufficient plasma temperature and a sufficient confinement quality, as measured by the product $n\tau$. At a temperature equivalent to 10,000 eV, the $n\tau$ product must be about 3×10^{20} particles per cubic metre times seconds.

Magnetic fusion energy generally creates plasmas with a density of about 3×10^{20} particles per cubic metre, which is about 10^{-8} of normal density. Hence, the characteristic time for heat to escape must be greater than about one second. This is a measure of the required degree of magnetic insulation for the heat content. Under these conditions the plasma remains in energy balance and can operate continuously if the ash of the nuclear fusion, namely helium, is removed (otherwise it will quench the plasma) and fuel is replenished.

ICF creates plasmas of much higher density, generally between 10^{31} and 10^{32} particles per cubic metre, or 1,000 to 10,000 times the normal density. As such, the confinement time, or minimum burn time, can be as short as 20×10^{-12} second. The objective in ICF is to achieve a temperature equivalent of 4,400 eV at the centre of the highly compressed fuel mass, while still having sufficient mass left around the centre so that the disassembly time will exceed the minimum burn time.

Fusion Energy Research

The fusion process has been studied in order to understand nuclear matter and forces, to learn more about the nuclear physics of stellar objects, and to develop thermonuclear weapons. During the late 1940s and early '50s, research programs in the United States, United Kingdom, and the Soviet Union began to yield a better understanding of nuclear fusion, and investigators embarked on ways of exploiting the process for practical energy production. Fusion reactor research focused primarily on using magnetic fields and electromagnetic forces to contain the extremely hot plasmas needed for thermonuclear fusion.

Researchers soon found, however, that it is exceedingly difficult to contain plasmas at fusion reaction temperatures because the hot gases tend to expand and escape from the enclosing magnetic structure. Plasma physics theory in the 1950s was incapable of describing the behavior of the plasmas in many of the early magnetic confinement systems.

The undeniable potential benefits of practical fusion energy led to an increasing call for international cooperation. American, British, and Soviet fusion programs were strictly classified until

1958, when most of their research programs were made public at the Second Geneva Conference on the Peaceful Uses of Atomic Energy, sponsored by the United Nations. Since that time, fusion research has been characterized by international collaboration. In addition, scientists have also continued to study and measure fusion reactions between the lighter elements so as to arrive at a more accurate determination of reaction rates. The formulas developed by nuclear physicists for predicting the rate of fusion energy generation have been adopted by astrophysicists to derive new information about the structure and evolution of stars.

Work on the other major approach to fusion energy, inertial confinement fusion (ICF), was begun in the early 1960s. The initial idea was proposed in 1961, only a year after the reported invention of the laser, in a then-classified proposal to employ large pulses of laser energy (which no one then quite knew how to achieve) to implode and shock-heat matter to temperatures at which nuclear fusion would proceed vigorously. Aspects of inertial confinement fusion were declassified in the 1970s and, especially, in the early 1990s to reveal important aspects of the design of the targets containing fusion fuels. Very painstaking and sophisticated work to design and develop short-pulse, high-power lasers and suitable millimetre-sized targets continues, and significant progress has been made.

Although practical fusion reactors have not been built yet, the necessary conditions of plasma temperature and heat insulation have been largely achieved, suggesting that fusion energy for electric-power production is now a serious possibility. Commercial fusion reactors promise an inexhaustible source of electricity for countries worldwide. From a practical viewpoint, however, the initiation of nuclear fusion in a hot plasma is but the first step in a whole sequence of steps required to convert fusion energy to electricity. In the end, successful fusion power systems must be capable of producing electricity safely and in a cost-effective manner, with a minimum of radioactive waste and environmental impact. The quest for practical fusion energy remains one of the great scientific and engineering challenges of humankind.

Neutron Generators

Neutron generators are neutron source devices which contain compact linear particle accelerators and that produce neutrons by fusing isotopes of hydrogen together. The fusion reactions take place in these devices by accelerating either deuterium, tritium, or a mixture of these two isotopes into a metal hydride target which also contains deuterium, tritium or a mixture of these isotopes. Fusion of deuterium atoms (D + D) results in the formation of a He^{-3} ion and a neutron with a kinetic energy of approximately 2.5 MeV. Fusion of a deuterium and a tritium atom (D + T) results in the formation of a He^{-4} ion and a neutron with a kinetic energy of approximately 14.1 MeV. Neutron generators have applications in medicine, security, and materials analysis.

The basic concept was first developed by Ernest Rutherford's team in the Cavendish Laboratory in the early 1930s. Using a linear accelerator driven by a Cockcroft–Walton generator, Mark Oliphant led an experiment that fired deuterium ions into a deuterium-infused metal foil and noticed that a small number of these particles gave off alpha particles. This was the first demonstration of nuclear fusion, as well as the first discovery of Helium-3 and tritium, created in these reactions. The introduction of new power sources has continually shrunk the size of these machines, from Oliphant's that filled the corner of the lab, to modern machines that are highly portable. Thousands of such small, relatively inexpensive systems have been built over the past five decades.

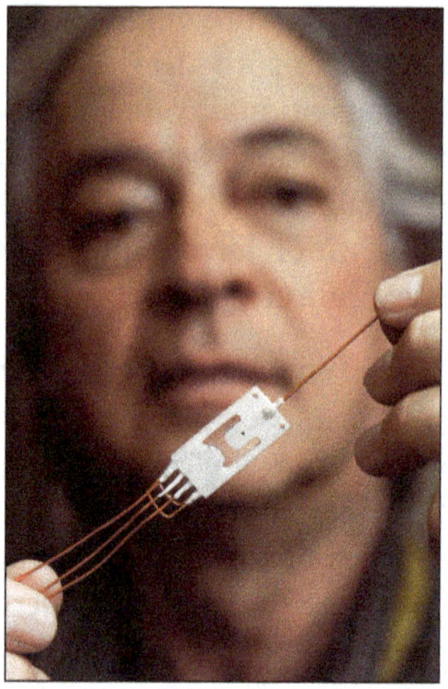

Neutristor in its simplest form as tested by the inventor at Sandia National Laboratories.

While neutron generators do produce fusion reactions, the number of accelerated ions that cause these reactions is very low. It can be easily demonstrated that the energy released by these reactions is many times lower than the energy needed to accelerate the ions, so there is no possibility of these machines being used to produce net fusion power. A related concept, colliding beam fusion, attempts to address this issue using two accelerators firing at each other.

Neutron Generator Theory and Operation

Small neutron generators using the deuterium (D, hydrogen-2, ^2H) tritium (T, hydrogen-3, ^3H) fusion reactions are the most common accelerator based (as opposed to radioactive isotopes) neutron sources. In these systems, neutrons are produced by creating ions of deuterium, tritium, or deuterium and tritium and accelerating these into a hydride target loaded with deuterium, deuterium, or deuterium and tritium. The DT reaction is used more than the DD reaction because the yield of the DT reaction is 50–100 times higher than that of the DD reaction.

$$D + T \rightarrow n + {}^4He \quad E_n = 14.1\ MeV$$

$$D + D \rightarrow n + {}^3He \quad E_n = 2.5\ MeV$$

Neutrons produced by DD and DT reactions are emitted somewhat anisotropically from the target, slightly biased in the forward (in the axis of the ion beam) direction. The anisotropy of the neutron emission from DD and DT reactions arises from the fact the reactions are isotropic in the center of momentum coordinate system (COM) but this isotropy is lost in the transformation from the COM coordinate system to the laboratory frame of reference. In both frames of reference, the He nuclei recoil in the opposite direction to the emitted neutron consistent with the law of conservation of momentum.

The gas pressure in the ion source region of the neutron tubes generally ranges between 0.1–0.01 mm Hg. The mean free path of electrons must be shorter than the discharge space to achieve ionization (lower limit for pressure) while the pressure must be kept low enough to avoid formation of discharges at the high extraction voltages applied between the electrodes. The pressure in the accelerating region, however, has to be much lower, as the mean free path of electrons must be longer to prevent formation of a discharge between the high voltage electrodes.

The ion accelerator usually consists of several electrodes with cylindrical symmetry, acting as an einzel lens. The ion beam can thus be focused to a small point at the target. The accelerators typically require power supplies of 100 - 500 kV. They usually have several stages, with voltage between the stages not exceeding 200 kV to prevent field emission.

In comparison with radionuclide neutron sources, neutron tubes can produce much higher neutron fluxes and consistent (monochromatic) neutron energy spectra can be obtained. The neutron production rate can also be controlled.

Sealed Neutron Tubes

The central part of a neutron generator is the particle accelerator itself, sometimes called a neutron tube. Neutron tubes have several components including an ion source, ion optic elements, and a beam target; all of these are enclosed within a vacuum-tight enclosure. High voltage insulation between the ion optical elements of the tube is provided by glass and ceramic insulators. The neutron tube is, in turn, enclosed in a metal housing, the accelerator head, which is filled with a dielectric medium to insulate the high voltage elements of the tube from the operating area. The accelerator and ion source high voltages are provided by external power supplies. The control console allows the operator to adjust the operating parameters of the neutron tube. The power supplies and control equipment are normally located within 10–30 feet of the accelerator head in laboratory instruments, but may be several kilometers away in well logging instruments.

In comparison with their predecessors, sealed neutron tubes do not require vacuum pumps and gas sources for operation. They are therefore more mobile and compact, while also durable and reliable. For example, sealed neutron tubes have replaced radioactive neutron initiators, in supplying a pulse of neutrons to the imploding core of modern nuclear weapons.

Examples of neutron tube ideas date as far back as the 1930s, pre-nuclear weapons era, by German scientists filing a 1938 German patent and obtaining a United States Patent; examples of present state of the art are given by developments such as the Neutristor, a mostly solid state device, resembling a computer chip, invented at Sandia National Laboratories in Albuquerque NM. Typical sealed designs are used in a pulsed mode and can be operated at different output levels, depending on the life from the ion source and loaded targets.

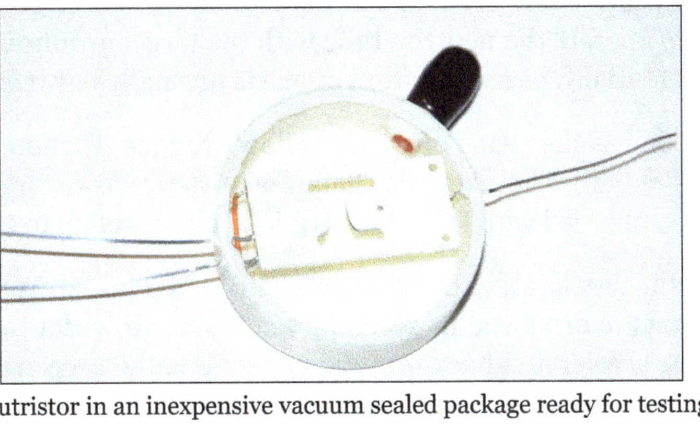

Neutristor in an inexpensive vacuum sealed package ready for testing.

Ion Sources

A good ion source should provide a strong ion beam without consuming much of the gas. For hydrogen isotopes, production of atomic ions is favored over molecular ions, as atomic ions have higher neutron yield on collision. The ions generated in the ion source are then extracted by an electric field into the accelerator region, and accelerated towards the target. The gas consumption is chiefly caused by the pressure difference between the ion generating and ion accelerating spaces that has to be maintained. Ion currents of 10 mA at gas consumptions of 40 cm^3/hour are achievable.

For a sealed neutron tube, the ideal ion source should use low gas pressure, give high ion current with large proportion of atomic ions, have low gas clean-up, use low power, have high reliability and high lifetime, its construction has to be simple and robust and its maintenance requirements have to be low.

Gas can be efficiently stored in a replenisher, an electrically heated coil of zirconium wire. Its temperature determines the rate of absorption/desorption of hydrogen by the metal, which regulates the pressure in the enclosure.

Cold Cathode (Penning)

The Penning source is a low gas pressure, cold cathode ion source which utilizes crossed electric and magnetic fields. The ion source anode is at a positive potential, either dc or pulsed, with respect to the source cathode. The ion source voltage is normally between 2 and 7 kilovolts. A magnetic field, oriented parallel to the source axis, is produced by a permanent magnet. A plasma is formed along the axis of the anode which traps electrons which, in turn, ionize gas in the source. The ions are extracted through the exit cathode. Under normal operation, the ion species produced by the Penning source are over 90% molecular ions. This disadvantage is however compensated for by the other advantages of the system.

One of the cathodes is a cup made of soft iron, enclosing most of the discharge space. The bottom of the cup has a hole through which most of the generated ions are ejected by the magnetic field into the acceleration space. The soft iron shields the acceleration space from the magnetic field, to prevent a breakdown.

Ions emerging from the exit cathode are accelerated through the potential difference between the exit cathode and the accelerator electrode. The schematic indicates that the exit cathode is at ground potential and the target is at high (negative) potential. This is the case in many sealed tube neutron generators. However, in cases when it is desired to deliver the maximum flux to a sample, it is desirable to operate the neutron tube with the target grounded and the source floating at high (positive) potential. The accelerator voltage is normally between 80 and 180 kilovolts.

The accelerating electrode has the shape of a long hollow cylinder. The ion beam has a slightly diverging angle (about 0.1 radian). The electrode shape and distance from target can be chosen so the entire target surface is bombarded with ions. Acceleration voltages of up to 200 kV are achievable.

The ions pass through the accelerating electrode and strike the target. When ions strike the target, 2–3 electrons per ion are produced by secondary emission. In order to prevent these secondary electrons from being accelerated back into the ion source, the accelerator electrode is biased negative with respect to the target. This voltage, called the suppressor voltage, must be at least

500 volts and may be as high as a few kilovolts. Loss of suppressor voltage will result in damage, possibly catastrophic, to the neutron tube.

Some neutron tubes incorporate an intermediate electrode, called the focus or extractor electrode, to control the size of the beam spot on the target. The gas pressure in the source is regulated by heating or cooling the gas reservoir element.

Radio Frequency (RF)

Ions can be created by electrons formed in high-frequency electromagnetic field. The discharge is formed in a tube located between electrodes, or inside a coil. Over 90% proportion of atomic ions is achievable.

Targets

The targets used in neutron generators are thin films of metal such as titanium, scandium, or zirconium which are deposited onto a silver, copper or molybdenum substrate. Titanium, scandium, and zirconium form stable chemical compounds called metal hydrides when combined with hydrogen or its isotopes. These metal hydrides are made up of two hydrogen (deuterium or tritium) atoms per metal atom and allow the target to have extremely high densities of hydrogen. This is important to maximize the neutron yield of the neutron tube. The gas reservoir element also uses metal hydrides, e.g. uranium hydride, as the active material.

Titanium is preferred to zirconium as it can withstand higher temperatures (200 °C), and gives higher neutron yield as it captures deuterons better than zirconium. The maximum temperature allowed for the target, above which hydrogen isotopes undergo desorption and escape the material, limits the ion current per surface unit of the target; slightly divergent beams are therefore used. A 1 microampere ion beam accelerated at 200 kV to a titanium-tritium target can generate up to 10^8 neutrons per second. The neutron yield is mostly determined by the accelerating voltage and the ion current level.

An example of a tritium target in use is a 0.2 mm thick silver disc with a 1 micrometer layer of titanium deposited on its surface; the titanium is then saturated with tritium.

Metals with sufficiently low hydrogen diffusion can be turned into deuterium targets by bombardment of deuterons until the metal is saturated. Gold targets under such condition show four times higher efficiency than titanium. Even better results can be achieved with targets made of a thin film of a high-absorption high-diffusivity metal (e.g. titanium) on a substrate with low hydrogen diffusivity (e.g. silver), as the hydrogen is then concentrated on the top layer and can not diffuse away into the bulk of the material. Using a deuterium-tritium gas mixture, self-replenishing D-T targets can be made. The neutron yield of such targets is lower than of tritium-saturated targets in deuteron beams, but their advantage is much longer lifetime and constant level of neutron production. Self-replenishing targets are also tolerant to high-temperature bake-out of the tubes, as their saturation with hydrogen isotopes is performed after the bakeout and tube sealing.

High Voltage Power Supplies

One particularly interesting approach for generating the high voltage fields needed to accelerate ions in a neutron tube is to use a pyroelectric crystal. In April 2005 researchers at UCLA demonstrated the use of a thermally cycled pyroelectric crystal to generate high electric fields in a neutron generator application. In February 2006 researchers at Rensselaer Polytechnic Institute demonstrated the use of two oppositely poled crystals for this application. Using these low-tech power supplies it is possible to generate a sufficiently high electric field gradient across an accelerating gap to accelerate deuterium ions into a deuterated target to produce the D + D fusion reaction. These devices are similar in their operating principle to conventional sealed-tube neutron generators which typically use Cockcroft–Walton type high voltage power supplies. The novelty of this approach is in the simplicity of the high voltage source. Unfortunately, the relatively low accelerating current that pyroelectric crystals can generate, together with the modest pulsing frequencies that can be achieved (a few cycles per minute) limits their near-term application in comparison with today's commercial products.

Neutron Radiation

Neutron radiation is a form of ionizing radiation that presents as free neutrons. Typical phenomena are nuclear fission or nuclear fusion causing the release of free neutrons, which then react with nuclei of other atoms to form new isotopes—which, in turn, may trigger further neutron radiation. Free neutrons are unstable, decaying into a proton, an electron, plus an anti-electron-neutrino with a mean lifetime of 887 seconds (about 14 minutes, 47 seconds).

Sources

Neutrons may be emitted from nuclear fusion or nuclear fission, or from other nuclear reactions such as radioactive decay or particle interactions with cosmic rays or within particle accelerators. Large neutron sources are rare, and usually limited to large-sized devices such as nuclear reactors or particle accelerators, including the Spallation Neutron Source.

Neutron radiation was discovered from observing an alpha particle colliding with a beryllium nucleus, which was transformed into a carbon nucleus while emitting a neutron, Be(α, n)C. The combination of an alpha particle emitter and an isotope with a large (α, n) nuclear reaction probability is still a common neutron source.

Neutron Radiation from Fission

The neutrons in nuclear reactors are generally categorized as slow (thermal) neutrons or fast neutrons depending on their energy. Thermal neutrons are similar in energy distribution (the Maxwell–Boltzmann distribution) to a gas in thermodynamic equilibrium but are easily captured by atomic nuclei and are the primary means by which elements undergo nuclear transmutation.

To achieve an effective fission chain reaction, neutrons produced during fission must be captured by fissionable nuclei, which then split, releasing more neutrons. In most fission reactor designs, the nuclear fuel is not sufficiently refined to absorb enough fast neutrons to carry on the chain reaction, due to the lower cross section for higher-energy neutrons, so a neutron moderator must be introduced to slow the fast neutrons down to thermal velocities to permit sufficient absorption.

Common neutron moderators include graphite, ordinary (light) water and heavy water. A few reactors (fast neutron reactors) and all nuclear weapons rely on fast neutrons. This requires certain changes in the design and in the required nuclear fuel. The element beryllium is particularly useful due to its ability to act as a neutron reflector or lens. This lets smaller quantities of fissile material be used and is a primary technical development that led to the creation of neutron bombs.

Cosmogenic Neutrons

Cosmogenic neutrons, neutrons produced from cosmic radiation in the Earth's atmosphere or surface, and those produced in particle accelerators can be significantly higher energy than those encountered in reactors. Most of them activate a nucleus before reaching the ground; a few react with nuclei in the air. The reactions with nitrogen-14 lead to the formation of carbon-14, widely used in radiocarbon dating.

Uses

Cold, thermal and hot neutron radiation is most commonly used in scattering and diffraction experiments, to assess the properties and the structure of materials in crystallography, condensed matter physics, biology, solid state chemistry, materials science, geology, mineralogy and related sciences. Neutron radiation is also used in Boron Neutron Capture Therapy to treat cancerous tumors due to its highly penetrating and damaging nature to cellular structure. Neutrons can also be used for imaging of industrial parts termed neutron radiography when using film, neutron radioscopy when taking a digital image, such as through image plates, and neutron tomography for three-dimensional images. Neutron imaging is commonly used in the nuclear industry, the space and aerospace industry, as well as the high reliability explosives industry.

Ionization Mechanisms and Properties

Neutron radiation is often called *indirectly ionizing radiation*. It does not ionize atoms in the same way that charged particles such as protons and electrons do (exciting an electron), because neutrons have no charge. However, neutron interactions are largely ionizing, for example when neutron absorption results in gamma emission and the gamma ray (photon) subsequently removes an electron from an atom, or a nucleus recoiling from a neutron interaction is ionized and causes more traditional subsequent ionization in other atoms. Because neutrons are uncharged, they are more penetrating than alpha radiation or beta radiation. In some cases they are more penetrating than gamma radiation, which is impeded in materials of high atomic number. In materials of low atomic number such as hydrogen, a low energy gamma ray may be more penetrating than a high energy neutron.

Health Hazards and Protection

In health physics neutron radiation is a type of radiation hazard. Another, sometimes more severe hazard of neutron radiation, is neutron activation, the ability of neutron radiation to induce radioactivity in most substances it encounters, including the body tissues. This occurs through the capture of neutrons by atomic nuclei, which are transformed to another nuclide, frequently a radionuclide. This process accounts for much of the radioactive material released by the detonation of a nuclear weapon. It is also a problem in nuclear fission and nuclear fusion installations as it gradually renders the equipment radioactive such that eventually it must be replaced and disposed of as low-level radioactive waste.

Neutron radiation protection relies on radiation shielding. Due to the high kinetic energy of neutrons, this radiation is considered the most severe and dangerous radiation to the whole body when it is exposed to external radiation sources. In comparison to conventional ionizing radiation based on photons or charged particles, neutrons are repeatedly bounced and slowed (absorbed) by light nuclei so hydrogen-rich material is more effective at shielding than iron nuclei. The light atoms serve to slow down the neutrons by elastic scattering so they can then be absorbed by nuclear reactions. However, gamma radiation is often produced in such reactions, so additional shielding must be provided to absorb it. Care must be taken to avoid using nuclei that undergo fission or neutron capture that causes radioactive decay of nuclei, producing gamma rays.

Neutrons readily pass through most material, and hence the absorbed dose (measured in Grays) from a given amount of radiation is low, but interact enough to cause biological damage. The most effective shielding materials are water, or hydrocarbons like polyethylene or paraffin wax. Water-extended polyester (WEP) is effective as a shielding wall in harsh environments due to its high hydrogen content and resistance to fire, allowing it to be used in a range of nuclear, health physics, and defense industries. Hydrogen-based materials are suitable for shielding as they are proper barriers against radiation.

Concrete (where a considerable number of water molecules chemically bind to the cement) and gravel provide a cheap solution due to their combined shielding of both gamma rays and neutrons. Boron is also an excellent neutron absorber (and also undergoes some neutron scattering). Boron decays into carbon or helium and produces virtually no gamma radiation with boron carbide, a shield commonly used where concrete would be cost prohibitive. Commercially, tanks of water or fuel oil, concrete, gravel, and B_4C are common shields that surround areas of large amounts of neutron flux, e.g., nuclear reactors. Boron-impregnated silica glass, standard borosilicate glass, high-boron steel, paraffin, and Plexiglas have niche uses.

Because neutrons that strike the hydrogen nucleus (proton, or deuteron) impart energy to that nucleus, they in turn break from their chemical bonds and travel a short distance before stopping. Such hydrogen nuclei are high linear energy transfer particles, and are in turn stopped by ionization of the material they travel through. Consequently, in living tissue, neutrons have a relatively high relative biological effectiveness, and are roughly ten times more effective at causing biological damage compared to gamma or beta radiation of equivalent energy exposure. These neutrons can either cause cells to change in their functionality or to completely stop reproducing, causing damage to the body over time. Neutrons are particularly damaging to soft tissues like the cornea of the eye.

Effects on Materials

High-energy neutrons damage and degrade materials over time; bombardment of materials with neutrons creates collision cascades that can produce point defects and dislocations in the material, the creation of which is the primary driver behind microstructural changes occurring over time in materials exposed to radiation. At high neutron fluences this can lead to embrittlement of metals and other materials, and to swelling in some of them. This poses a problem for nuclear reactor vessels and significantly limits their lifetime (which can be somewhat prolonged by controlled annealing of the vessel, reducing the number of the built-up dislocations). Graphite moderator blocks are especially susceptible to this effect, known as Wigner effect, and must be annealed periodically. The Windscale fire was caused by a mishap during such an annealing operation.

Radiation damage to materials occurs as a result of the interaction of an energetic incident particle (a neutron, or otherwise) with a lattice atom in the material. The collision causes a massive transfer of kinetic energy to the lattice atom, which is displaced from its lattice site, becoming what is known as the primary knock-on atom (PKA). Because the PKA is surrounded by other lattice atoms, its displacement and passage through the lattice results in many subsequent collisions and the creations of additional knock-on atoms, producing what is known as the collision cascade or displacement cascade. The knock-on atoms lose energy with each collision, and terminate as interstitials, effectively creating a series of Frenkel defects in the lattice. Heat is also created as a result of the collisions (from electronic energy loss), as are possibly transmuted atoms. The magnitude of the damage is such that a single 1 MeV neutron creating a PKA in an iron lattice produces approximately 1100 Frenkel pairs. The entire cascade event occurs over a timescale of 10^-13 seconds, and therefore, can only be 'observed' in computer simulations of the event.

The knock-on atoms terminate in non-equilibrium interstitial lattice positions, many of which annihilate themselves by diffusing back into neighboring vacant lattice sites and restore the ordered lattice. Those that do not or cannot leave vacancies, which causes a local rise in the vacancy concentration far above that of the equilibrium concentration. These vacancies tend to migrate as a result of thermal diffusion towards vacancy sinks (i.e., grain boundaries, dislocations) but exist for significant amounts of time, during which additional high-energy particles bombard the lattice, creating collision cascades and additional vacancies, which migrate towards sinks. The main effect of irradiation in a lattice is the significant and persistent flux of defects to sinks in what is known as the defect wind. Vacancies can also annihilate by combining with one another to form dislocation loops and later, lattice voids.

The collision cascade creates many more vacancies and interstitials in the material than equilibrium for a given temperature, and diffusivity in the material is dramatically increased as a result. This leads to an effect called radiation enhanced diffusion, which leads to microstructural evolution of the material over time. The mechanisms leading to the evolution of the microstructure are many, may vary with temperature, flux, and fluence, and are a subject of extensive study:

- Radiation-induced segregation results from the aforementioned flux of vacancies to sinks, implying a flux of lattice atoms away from sinks: but not necessarily in the same proportion to alloy composition in the case of an alloyed material. These fluxes may therefore lead to depletion of alloying elements in the vicinity of sinks. For the flux of interstitials introduced by the cascade, the effect is reversed: the interstitials diffuse toward sinks resulting in alloy enrichment near the sink.

- Dislocation loops are formed if vacancies form clusters on a lattice plane. If these vacancy concentration expand in three dimensions, a void forms. By definition, voids are under vacuum, but may became gas-filled in the case of alpha-particle radiation (helium) or if the gas is produced as a result of transmutation reactions. The void is then called a bubble, and leads to dimensional instability (swelling) of parts subject to radiation. Swelling presents a major long-term design problem, especially in reactor components made out of stainless steel. Alloys with crystallographic isotropy, such as Zircaloys are subject to the creation of dislocation loops, but do not exhibit void formation. Instead, the loops form on particular lattice planes, and can lead to irradiation-induced growth, a phenomenon distinct from swelling, but that can also produce significant dimensional changes in an alloy.

- Irradiation of materials can also induce phase transformations in the material: in the case of a solid solution, the solute enrichment or depletion at sinks radiation-induced segregation can lead to the precipitation of new phases in the material.

The mechanical effects of these mechanisms include irradiation hardening, embrittlement, creep, and environmentally-assisted cracking. The defect clusters, dislocation loops, voids, bubbles, and precipitates produced as a result of radiation in a material all contribute to the strengthening and embrittlement (loss of ductility) in the material Embrittlement is of particular concern for the material comprising the reactor pressure vessel, where as a result the energy required to fracture the vessel decreases significantly. It is possible to restore ductility by annealing the defects out, and much of the life-extension of nuclear reactors depends on the ability to safely do so. Creep is also greatly accelerated in irradiated materials, though not as a result of the enhanced diffusivities, but rather as a result of the interaction between lattice stress and the developing microstructure. Environmentally-assisted cracking or, more specifically, irradiation assisted stress corrosion cracking (IASCC) is observed especially in alloys subject to neutron radiation and in contact with water, caused by hydrogen absorption at crack tips resulting from radiolysis of the water, leading to a reduction in the required energy to propagate the crack.

Radioactive Decay

Radioactive decay is the process by which an excited, unstable atomic nucleus loses energy by emitting radiation in the form of particles or electromagnetic waves, thereby transitioning toward a more stable state.

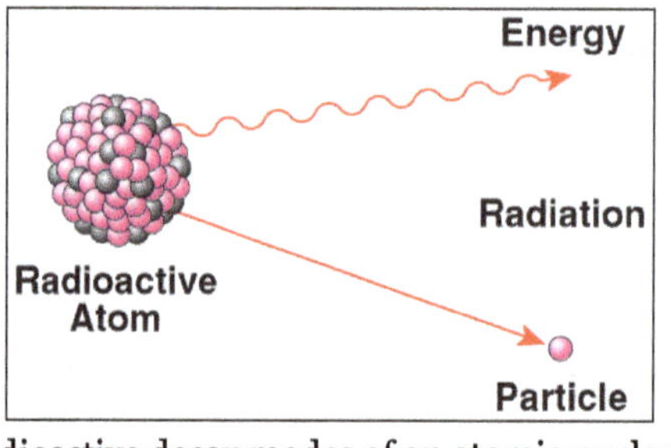

Radioactive decay modes of an atomic nucleus.

The atomic nucleus comprises certain combinations of protons and neutrons held in a stable configuration through a precise balance of powerful forces: The strong force holding the protons and neutrons together is powerful but very short range; the electrostatic repulsion of the positively charged protons is less powerful but long range; the weak force makes the neutron inherently unstable and will turn it into a proton if given the chance. This balance is very delicate: a uranium-238 nucleus has a half-life of 4.5 billion years while uranium-237 with just one less neutron has a half-life of 1.3 minutes.

If there is an imbalance in these forces, the system will eventually shed the excess by ejecting radiation in some combination of particles and wave energy. The most common radioactive decays occur in response to one of three possible types of imbalance. If the nucleus has too many neutrons, one of its neutrons decays (through beta decay) into one proton plus two fragments ejected from the nucleus, a neutrino and an electron (called a a beta particle). If the nucleus has too many protons, it undergoes alpha decay by ejecting two protons and two neutrons as an alpha particle. If the nucleus is excited (has too much energy) it ejects a gamma ray.

Materials exhibiting radioactive decay have yielded widespread application to enhance human welfare. The various applications take advantage of the different decay properties, different decay products, and different chemical properties of the many elements having some isotopes that are radioactive. Major types of applications use the radiation either for diagnosing a problem or for treating a problem by killing specific harmful cells. Areas of application include human and veterinary medicine, nutrition research, basic research in genetics and metabolism, household smoke detectors, industrial and mining inspection of welds, security inspection of cargo, tracing and analyzing pollutants in studies of runoff, and dating materials in geology, paleontology, and archaeology.

Nuclides

Radioactive decay results in an atom of one type, called the parent nuclide, being transformed to an atom of a different type, called the daughter nuclide. For example, a carbon-14 atom (the "parent") emits radiation and transforms to a nitrogen-14 atom (the "daughter"). This transformation involves quantum probability, so it is impossible to predict when a particular atom will decay. Given a large number of atoms, however, the decay rate is predictable and measured by the "half-life"—the time it takes for 50 percent of the atoms to undergo the change. The half-life of radioactive atoms varies enormously; from fractions of a millisecond to billions of years.

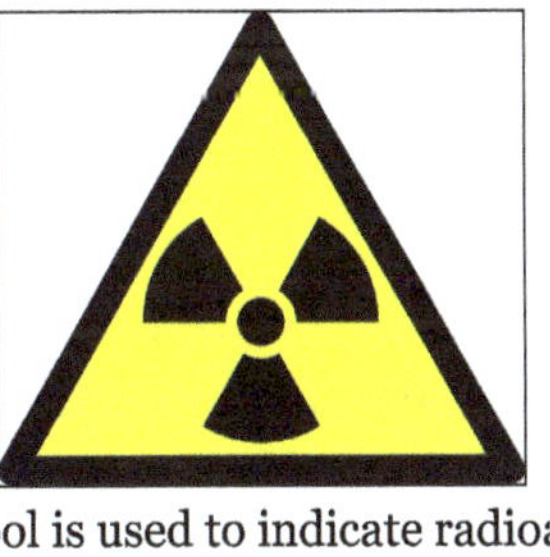

The trefoil symbol is used to indicate radioactive material.

The SI unit of radioactive decay (the phenomenon of natural and artificial radioactivity) is the becquerel (Bq). One Bq is defined as one transformation (or decay) per second. Since any reasonably-sized sample of radioactive material contains many atoms, a Bq is a tiny measure of activity; amounts on the order of TBq (terabecquerel) or GBq (gigabecquerel) are commonly used. Another unit of (radio)activity is the curie, Ci, which was originally defined as the activity of one gram of pure radium, isotope Ra-226. At present, it is equal (by definition) to the activity of any radionuclide decaying with a disintegration rate of 3.7×10^{10} Bq. The use of Ci is presently discouraged by SI.

Explanation

The neutrons and protons that constitute nuclei, as well as other particles that may approach them, are governed by several interactions. The strong nuclear force, not observed at the familiar macroscopic scale, is the most powerful force over subatomic distances. The electrostatic force is also significant, while the weak nuclear force is responsible for Beta decay.

The interplay of these forces is simple. Some configurations of the particles in a nucleus have the property that, should they shift ever so slightly, the particles could fall into a lower-energy arrangement (with the extra energy moving elsewhere). One might draw an analogy with a snowfield on a mountain: While friction between the snow crystals can support the snow's weight, the system is inherently unstable with regards to a lower-potential-energy state, and a disturbance may facilitate the path to a greater entropy state (that is, towards the ground state where heat will be produced, and thus total energy is distributed over a larger number of quantum states). Thus, an avalanche results. The total energy does not change in this process, but because of entropy effects, avalanches only happen in one direction, and the end of this direction, which is dictated by the largest number of chance-mediated ways to distribute available energy, is what we commonly refer to as the "ground state".

Such a collapse (a decay event) requires a specific activation energy. In the case of a snow avalanche, this energy classically comes as a disturbance from outside the system, although such disturbances can be arbitrarily small. In the case of an excited atomic nucleus, the arbitrarily small disturbance comes from quantum vacuum fluctuations. A nucleus (or any excited system in quantum mechanics) is unstable, and can thus spontaneously stabilize to a less-excited system. This process is driven by entropy considerations: The energy does not change, but at the end of the process, the total energy is more diffused in spacial volume. The resulting transformation alters the structure of the nucleus. Such a reaction is thus a nuclear reaction, in contrast to chemical reactions, which also are driven by entropy, but which involve changes in the arrangement of the outer electrons of atoms, rather than their nuclei.

Some nuclear reactions do involve external sources of energy, in the form of collisions with outside particles. However, these are not considered decay. Rather, they are examples of induced nuclear reactions. Nuclear fission and fusion are common types of induced nuclear reactions.

Modes of Decay

Radionuclides can undergo a number of different reactions. These are summarized in the following table. A nucleus with atomic weight A and a positive charge Z (called atomic number) is represented as (A, Z).

Mode of decay	Participating particles	Daughter nucleus
Decays with emission of nucleons:		
Alpha decay	An alpha particle (A=4, Z=2) emitted from nucleus	(A-4, Z-2)
Proton emission	A proton ejected from nucleus	(A-1, Z-1)
Neutron emission	A neutron ejected from nucleus	(A-1, Z)
Double proton emission	Two protons ejected from nucleus simultaneously	(A-2, Z-2)
Spontaneous fission	Nucleus disintegrates into two or more smaller nuclei and other particles	-
Cluster decay	Nucleus emits a specific type of smaller nucleus (A_1, Z_1) larger than an alpha particle	(A-A_1, Z-Z_1) + (A_1,Z_1)

Different modes of beta decay:		
Beta-Negative decay	A nucleus emits an electron and an antineutrino	(*A*, *Z*+1)
Positron emission, also Beta-Positive decay	A nucleus emits a positron and a neutrino	(*A*, *Z*-1)
Electron capture	A nucleus captures an orbiting electron and emits a neutrino - The daughter nucleus is left in an excited and unstable state	(*A*, *Z*-1)
Double beta decay	A nucleus emits two electrons and two antineutrinos	(*A*, *Z*+2)
Double electron capture	A nucleus absorbs two orbital electrons and emits two neutrinos - The daughter nucleus is left in an excited and unstable state	(*A*, *Z*-2)
Electron capture with positron emission	A nucleus absorbs one orbital electron, emits one positron and two neutrinos	(*A*, *Z*-2)
Double positron cmission	A nucleus emits two positrons and two neutrinos	(*A*, *Z*-2)
Transitions between states of the same nucleus:		
Gamma decay	Excited nucleus releases a high-energy photon (gamma ray)	(*A*, *Z*)
Internal conversion	Excited nucleus transfers energy to an orbital electron and it is ejected from the atom	(*A*, *Z*)

Radioactive decay results in a reduction of summed rest mass, which is converted to energy (the disintegration energy) according to the formula $E = mc^2$. This energy is released as kinetic energy of the emitted particles. The energy remains associated with a measure of mass of the decay system invariant mass, inasmuch the kinetic energy of emitted particles contributes also to the total invariant mass of systems. Thus, the sum of rest masses of particles is not conserved in decay, but the system mass or system invariant mass (as also system total energy) is conserved.

Radioactive Series

In a simple, one-step radioactive decay, the new nucleus that emerges is stable. C-14 undergoing beta decay to N-14 and K-40 undergoing electron capture to Ar-40 are examples.

On the other hand, the daughter nuclide of a decay event can be unstable, sometimes even more unstable than the parent. If this is the case, it will proceed to decay again. A sequence of several decay events, producing in the end a stable nuclide, is a decay chain. Ultrapure uranium, for instance, is hardly radioactive at all. After a few weeks, however, the unstable daughter nucleides accumulate—such as radium—and it is their radioactivity that becomes noticeable. Of the commonly occurring forms of radioactive decay, the only one that changes the number of aggregate protons and neutrons (nucleons) contained in the nucleus is alpha emission, which reduces it by four. Thus, the number of nucleons modulo 4 is preserved across any decay chain. This leads to the four radioactive decay series with atomic weights 4n+0, 4n+1, 4n+2, and 4n+3.

In an alpha decay, the atomic weight decreases by 4 and the atomic number decreases by 2. In a beta decay, the atomic weight stays the same and the atomic number increases by 1. In a gamma decay, both atomic weight and number stay the same. A branching path occurs when there are alternate routes to the same stable destination. One branch is usually highly favored over the other.

These are the four radioactive decay series.

Uranium-235 Series (4n+3)

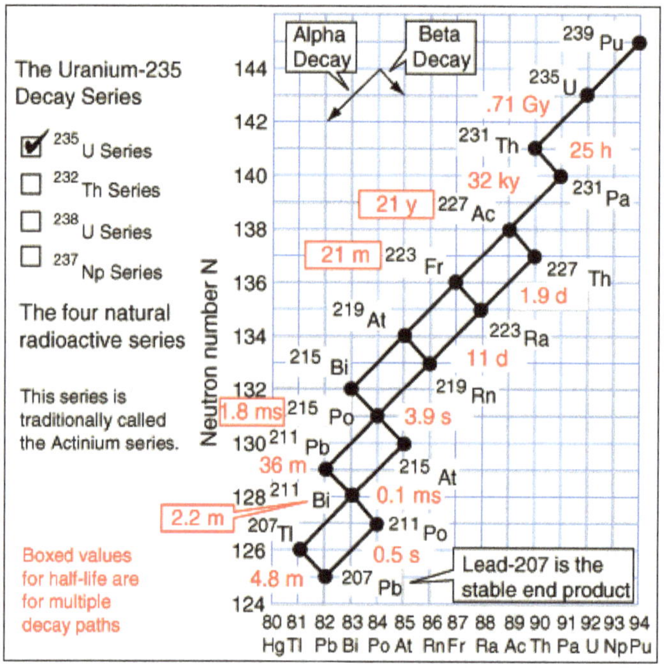

Thorium-232 Series (4n+0)

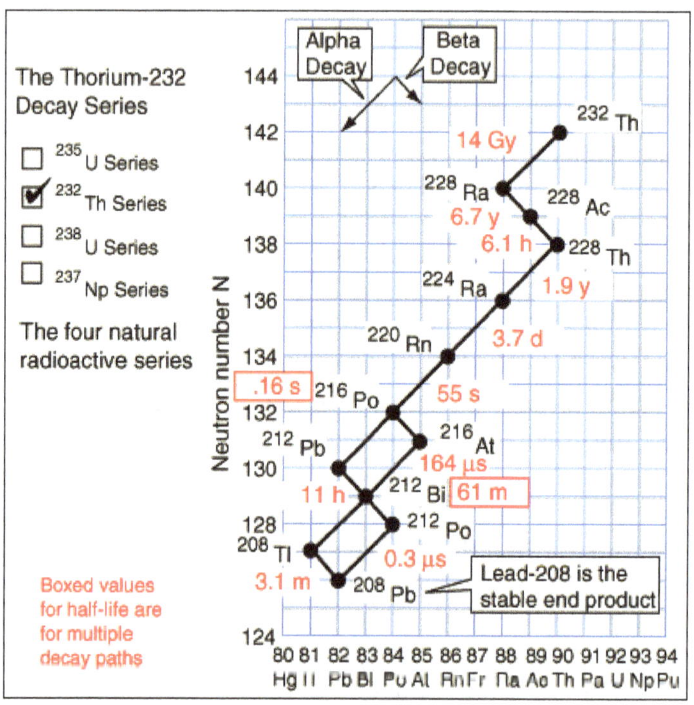

Uranium-238 Series (4n+2)

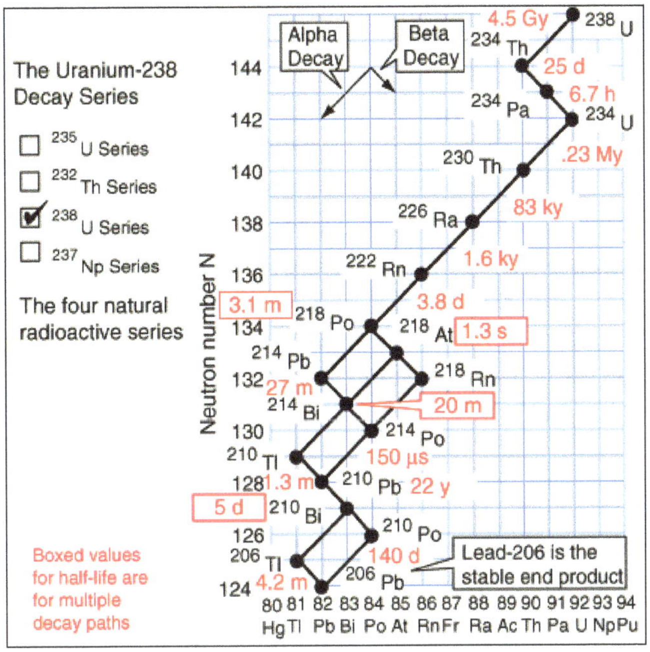

Neptunium-237 Series (4n+1)

The members of this series are not presently found in nature because the half-life of the longest lived isotope in the series is short compared to the age of the earth.

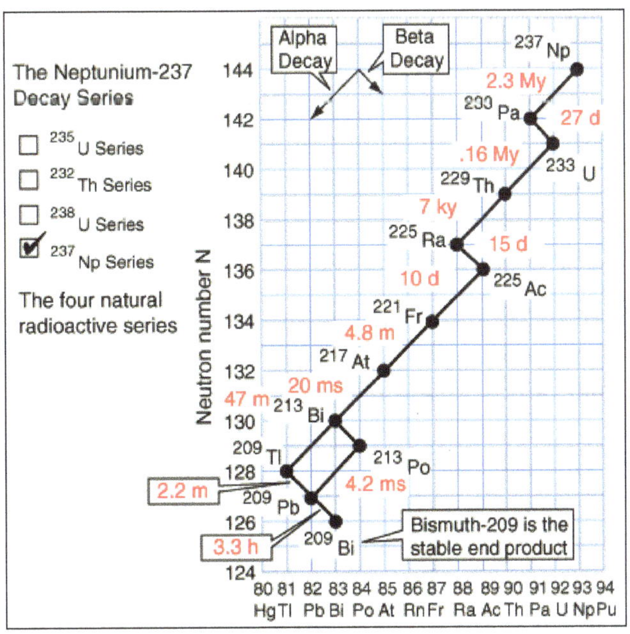

Occurrence

According to the widely accepted Big Bang theory, the universe began as a mixture of hydrogen-1 (75 percent) and helium-4 (25 percent) with only traces of other light atoms. All the other elements, including the radioactive ones, were generated later during the thermonuclear burning of stars—the fusion of the lighter elements into the heavier ones. Stable isotopes of the lightest five elements (H, He, and traces of Li, Be, and B) were produced very shortly after the emergence of the universe, in a process called Big Bang nucleosynthesis. These lightest stable nuclides (including deuterium) survive to today, but any radioactive isotopes of the light elements produced in the Big Bang (such as tritium) have long since decayed. Isotopes of elements heavier than boron were not produced at all in the Big Bang, and these first five elements do not have any long-lived radioisotopes. Thus, all radioactive nuclei are, therefore, relatively young with respect to the birth of the universe, having formed later in various other types of nucleosynthesis in stars (in particular, supernovae), and also during ongoing interactions between stable isotopes and energetic particles. For example, carbon-14, a radioactive nuclide with a half-life of only 5,730 years, is constantly produced in Earth's upper atmosphere due to interactions between cosmic rays and nitrogen.

Applications

Radioactive materials and their decay products—alpha particles (2 protons plus 2 neutrons), beta particles (electrons or positrons), gamma radiation, and the daughter isotopes—have been put to the service of humanity in a great number of ways. At the same time, high doses of radiation from radioactive materials can be toxic unless they are applied with medical precision and control. Such exposures are unlikely except for the unlikely cases of a nuclear weapon detonation or an accident or attack on a nuclear facility.

In medicine, some radioactive isotopes, such as iron-59 and iodine-131, are usable directly in the body because the isotopes are chemically the same as stable iron and iodine respectively. Iron-59, steadily announcing its location by emitting beta-decay electrons, is readily incorporated into blood cells and thereby serves as an aid in studying iron deficiency, a nutritional deficiency affecting more than 2 billion people globally. Iron-59 is an important tool in the effort to understand the many factors affecting a person's ability to metabolize iron in the diet so that it becomes part of the the blood. Iodine-131 administered in the blood to people suffering from hyperthyroidism or thyroid cancer concentrates in the thyroid where gamma radiation emitted by the iodine-131 kills many of the thyroid cells. Hyperthyroidism in cats is treated effectively by one dose of iodine-131.

Radioactive isotopes whose chemical nature does not permit them to be readily incorporated into the body, are delivered to targeted areas by attaching them to a particular molecule that does tend to concentrate in a particular bodily location—just as iodine naturally concentrates in the thyroid gland. For studying activity in the brain, the radioactive isotope fluorine-18 is commonly attached to an analog of the sugar glucose which tends to concentrate in the active regions of the brain within a short time after the molecule is injected into the blood. Fluorine-18 decays by releasing a positron whose life is soon ended as it meets an electron and the two annihilate yielding gamma radiation that is readily detected by the Positron Emission Tomography (PET) technology. Similar techniques of radioisotopic labeling, have been used to track the passage of a variety of chemical substances through complex systems, especially living organisms.

Three gamma emitting radioisotopes are commonly used as a source of radiation. Technetium-99m, a metastable form with a half-life of 6 hours, emits a relatively low frequency gamma radiation that is readily detected. It has been widely used for imaging and functional studies of the brain, myocardium, thyroid, lungs, liver, gallbladder, kidneys, skeleton, blood, and tumors. Gamma radiation from cobalt-60 is used for sterilizing medical equipment, treating cancer, pasteurizing certain foods and spices, gauging the thickness of steel as it is being produced, and monitoring welds. Cesium-137 is used as a source of gamma radiation for treating cancer, measuring soil density at construction sites, monitoring the filling of packages of foods and pharmaceuticals, monitoring fluid flows in production plants, and studying rock layers in oil wells.

Americanium-241, which decays by emitting alpha particles and low energy gamma radiation, is commonly used in smoke detectors as the alpha particles ionize air in a chamber permitting a small current to flow. Smoke particles entering the chamber activate the detector by absorbing alpha particles without being ionized, thereby reducing the current.

On the premise that radioactive decay is truly random (rather than merely chaotic), it has been used in hardware random-number generators. Because the process is not thought to vary significantly in mechanism over time, it is also a valuable tool in estimating the absolute ages of certain materials. For geological materials, the radioisotopes (parents) and certain of their decay products (daughters) become trapped when a rock solidifies, and can then later be used to estimate the date of the solidification (subject to such uncertainties as the possible number of daughter elements present at the time of solidification and the possible number of parent or daughter atoms added or removed over time).

For dating organic matter, radioactive carbon-14 is used because the atmosphere contains a small percentage of carbon-14 along with the predominance of stable carbons 12 and 13. Living plants incorporate the same ratio of carbon-14 to carbon-12 into their tissues and the animals eating the plants have a similar ratio in their tissues. After organisms die, their carbon-14 decays to nitrogen at a certain rate while the carbon-12 content remains constant. Thus, in principle, measuring the ratio of carbon-14 to carbon-12 in the dead organism provides an indication of how long the organism has been dead. This dating method is limited by the 5730 year half-life of carbon-14 to a maximum of 50,000 to 60,000 years. The accuracy of carbon dating has been called into question primarily because the concentration of carbon-14 in the atmosphere varies over time and some plants have the capacity to exclude carbon-14 from their intake.

Radioactive Decay Rates

The decay rate, or activity, of a radioactive substance are characterized by:

- Constant Quantities:
    - Half life: Symbol $t_{1/2}$ —the time for half of a substance to decay.
    - Mean lifetime: Symbol ta—the average lifetime of any given particle.
    - Decay constant: Symbol —the inverse of the mean lifetime.

(Note that although these are constants, they are associated with statistically random behavior of substances, and predictions using these constants are less accurate for a small number of atoms).

- Time-variable Quantities:
    - Total activity: Symbol A—number of decays an object undergoes per second.

- Number of particles: Symbol N—the total number of particles in the sample.
- Specific activity: Symbol S_A—number of decays per second per amount of substance. The "amount of substance" can be the unit of either mass or volume.

These are related as follows:

$$t_{1/2} = \frac{\ln(2)}{\lambda} = \tau \ln(2)$$

$$A = -\frac{dN}{dt} = \quad N$$

$$S_A a_0 = -\frac{dN}{dt}\Big|_{t=0} = \lambda N_0$$

Where, a_0 is the initial amount of active substance—substance that has the same percentage of unstable particles as when the substance was formed.

Activity Measurements

The units in which activities are measured are: Becquerel (symbol Bq) = number of disintegrations per second; curie (Ci) = 3.7×10^{10} disintegrations per second. Low activities are also measured in disintegrations per minute (dpm).

Decay Timing

The decay of an unstable nucleus is entirely random and it is impossible to predict when a particular atom will decay. However, it is equally likely to decay at any time. Therefore, given a sample of a particular radioisotope, the number of decay events $-dN$ expected to occur in a small interval of time dt is proportional to the number of atoms present. If N is the number of atoms, then the probability of decay ($-dN/N$) is proportional to dt:

$$\left(-\frac{dN}{N}\right) = \lambda \cdot dt$$

Particular radionuclides decay at different rates, each having its own decay constant (λ). The negative sign indicates that N decreases with each decay event. The solution to this first-order differential equation is the following function:

$$N(t) = N_0 e^{-\lambda t} = N_0 e^{-t/\tau}$$

This function represents exponential decay. It is only an approximate solution, for two reasons. Firstly, the exponential function is continuous, but the physical quantity N can only take non-negative integer values. Secondly, because it describes a random process, it is only statistically true. However, in most common cases, N is a very large number and the function is a good approximation.

In addition to the decay constant, radioactive decay is sometimes characterized by the mean life-time. Each atom "lives" for a finite amount of time before it decays, and the mean lifetime is the arithmetic mean of all the atoms' lifetimes. It is represented by the symbol \tau, and is related to the decay constant as follows:

$$\tau = \frac{1}{\lambda}$$

A more commonly used parameter is the half-life. Given a sample of a particular radionuclide, the half-life is the time taken for half the radionuclide's atoms to decay. The half life is related to the decay constant as follows:

$$t_{1/2} = \frac{\ln 2}{\lambda} = \tau \ln 2$$

This relationship between the half-life and the decay constant shows that highly radioactive substances are quickly spent, while those that radiate weakly endure longer. Half-lives of known radionuclides vary widely, from more than 10^{19} years (such as for very nearly stable nuclides, for example, ^{209}Bi), to 10^{-23} seconds for highly unstable ones.

References

- Nuclear-reactions, nuclear-power: nuclear-power.net, Retrieved 2 June, 2019
- Vichi, Sara (2016). "Efficiency calibration of a portable CZT detector for". Radiation Effects and Defects in Solids. 171: 705–713. Doi:10.1080/10420150.2016.1244675
- Nuclear-binding-energy-and-mass-defect, introchem: lumenlearning.com, Retrieved 3 July, 2019
- Bendjebbar, André (2000). Histoire secrète de la bombe atomique française. Documents (in French). Paris. Cherche Midi. ISBN 978-2-862-74794-1. OCLC 45842105
- Nuclear-fusion, science: britannica.com, Retrieved 3 July, 2019
- Jogalekar, Ashutosh. "Leo Szil rd, a traffic light and a slice of nuclear history". Scientific American. Retrieved 4 January2016
- Radioactive-decay, entry: newworldencyclopedia.org, Retrieved 4 August, 2019
- Parekh, PP; Semkow, TM; Torres, MA; Haines, DK; Cooper, JM; Rosenberga, PM; Kittoa, ME (2006). "Radioactivity in Trinitite six decades later". Journal of Environmental Radioactivity. 85 (1): 103–120. Doi:10.1016/j.jenvrad.2005.01.017. PMID 16102878

4

Components of a Nuclear Power Plant

Some of the basic components of nuclear power plants are nuclear reactors, steam generators, reactor coolant pumps, cooling towers, spent fuel pool and containment building. The device which is used to initiate and control a self-sustained nuclear chain reaction is known as a nuclear reactor. The topics elaborated in this chapter will help in gaining a better perspective about these components of a nuclear power plant.

Nuclear power plants are a type of power plant that use the process of nuclear fission in order to generate electricity. They do this by using nuclear reactors in combination with the Rankine cycle, where the heat generated by the reactor converts water into steam, which spins a turbine and a generator. Nuclear power provides the world with around 11% of its total electricity, with the largest producers being the United States and France.

The Darlington nuclear power plant in Ontario produces power from four 878 MW CANDU reactors.

Aside from the source of heat, nuclear power plants are very similar to coal-fired power plants. However, they require different safety measures since the use of nuclear fuel has vastly different properties from coal or other fossil fuels. They get their thermal power from splitting the nuclei of atoms in their reactor core, with uranium being the dominant choice of fuel in the world today. Thorium also has potential use in nuclear power production, however it is not currently in use. Below is the basic operation of a boiling water power plant, which shows the many components of a power plant, along with the generation of electricity.

Components and Operation

Nuclear Reactor

The reactor is a key component of a power plant, as it contains the fuel and its nuclear chain reaction, along with all of the nuclear waste products. The reactor is the heat source for the power plant, just like the boiler is for a coal plant. Uranium is the dominant nuclear fuel used in nuclear reactors, and its fission reactions are what produce the heat within a reactor. This heat is then transferred to the reactor's coolant, which provides heat to other parts of the nuclear power plant.

Besides their use in power generation, there are other types of nuclear reactors that are used for plutonium manufacturing, the propulsion of ships, aircraft and satellites, along with research and medical purposes. The power plant encompasses not just the reactor, but also cooling towers, turbines, generators, and various safety systems. The reactor is what makes it differ from other external heat engines.

Steam Generation

The production of steam is common among all nuclear power plants, but the way this is done varies immensely.

Steam turbine in a power plant.

The most common power plants in the world use pressurized water reactors, which use two loops of circling water to produce steam. The first loop carries extremely hot liquid water to a heat exchanger, where water at a lower pressure is circulated. It then heats up and boils to steam, and can then be sent to the turbine.

Boiling water reactors, the second most common reactor in power generation, heat the water in the core directly to steam.

Turbine and Generator

Once steam has been produced, it travels at high pressures and speeds through one or more turbines. These get up to extremely high speeds, causing the steam to loose energy, therefore, condensing back to a cooler liquid water. The rotation of the turbines is used to spin an electric generator, which produces electricity that is sent out the the electrical grid.

Two cooling towers of a nuclear power plant.

Cooling Towers

Perhaps the most iconic symbol of a nuclear power plant is the cooling towers. They work to reject waste heat to the atmosphere by the transfer of heat from hot water to the cooler outside air. Hot water cools in contact with the air and a small portion, around 2%, evaporates and raises up through the top. Moreover, these plants do not release any carbon dioxide—the primary greenhouse gas that contributes to climate change.

Many nuclear power plants simply put the waste heat into a river, lake or ocean instead of having cooling towers. Many other power plants like coal-fired power plants have cooling towers or these large bodies of water as well. This similarity exists because the process of turning heat into electricity is almost identical between nuclear power plants and coal-fired power plants.

Efficiency

The efficiency of a nuclear power plant is determined similarly to other heat engines—since technically the plant is a large heat engine. The amount of electric power produced for each unit of thermal power gives the plant its thermal efficiency, and due to the second law of thermodynamics there is an upper limit to how efficient these plants can be.

Typical nuclear power plants achieve efficiencies around 33-37%, comparable to fossil fueled power plants. Higher temperature and more modern designs like the Generation IV nuclear reactors could potentially reach above 45% efficiency.

Nuclear Reactor

Nuclear reactor is any of a class of devices that can initiate and control a self-sustaining series of nuclear fissions. Nuclear reactors are used as research tools, as systems for producing radioactive isotopes, and most prominently as energy sources for nuclear power plants.

Principles of Operation

Nuclear reactors operate on the principle of nuclear fission, the process in which a heavy atomic nucleus splits into two smaller fragments. The nuclear fragments are in very excited states and emit neutrons, other subatomic particles, and photons. The emitted neutrons may then cause new fissions, which in turn yield more neutrons, and so forth. Such a continuous self-sustaining series of fissions constitutes a fission chain reaction. A large amount of energy is released in this process, and this energy is the basis of nuclear power systems.

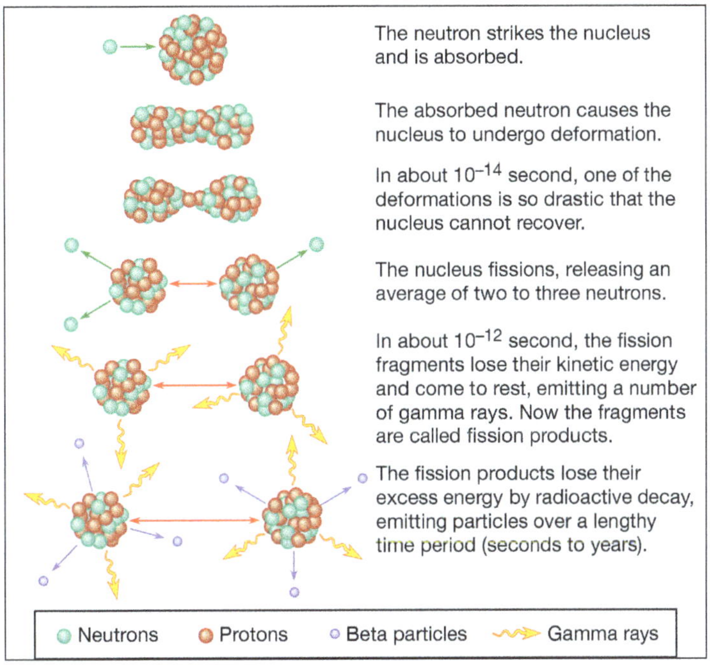

Sequence of events in the fission of a uranium nucleus by a neutron.

In an atomic bomb the chain reaction is designed to increase in intensity until much of the material has fissioned. This increase is very rapid and produces the extremely prompt, tremendously energetic explosions characteristic of such bombs. In a nuclear reactor the chain reaction is maintained at a controlled, nearly constant level. Nuclear reactors are so designed that they cannot explode like atomic bombs.

Most of the energy of fission—approximately 85 percent of it—is released within a very short time after the process has occurred. The remainder of the energy produced as a result of a fission event comes from the radioactive decay of fission products, which are fission fragments after they have emitted neutrons. Radioactive decay is the process by which an atom reaches a more stable state; the decay process continues even after fissioning has ceased, and its energy must be dealt with in any proper reactor design.

Chain Reaction and Criticality

The course of a chain reaction is determined by the probability that a neutron released in fission will cause a subsequent fission. If the neutron population in a reactor decreases over a given period of time, the rate of fission will decrease and ultimately drop to zero. In this case the reactor will be in what is known as a subcritical state. If over the course of time the neutron population is sustained at a constant rate, the fission rate will remain steady, and the reactor will be in what is called a critical state. Finally, if the neutron population increases over time, the fission rate and power will increase, and the reactor will be in a supercritical state.

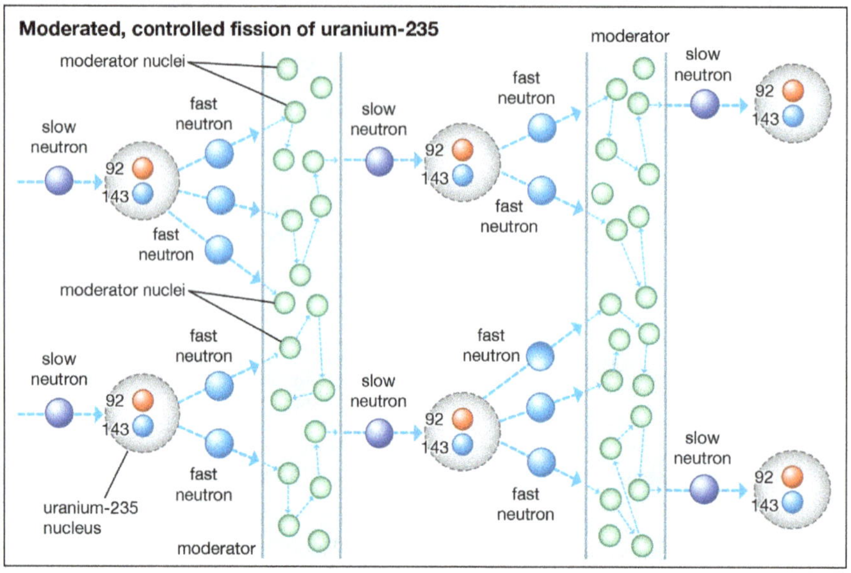

Chain reaction in a nuclear reactor at a critical state.

Chain reaction in a nuclear reactor at a critical stateSlow neutrons strike nuclei of uranium-235, causing the nuclei to fission, or split, and release fast neutrons. The fast neutrons are absorbed or slowed by the nuclei of a graphite moderator, which allows just enough slow neutrons to continue the fission chain reaction at a constant rate.

Before a reactor is started up, the neutron population is near zero. During reactor start-up, operators remove control rods from the core in order to promote fissioning in the reactor core, effectively putting the reactor temporarily into a supercritical state. When the reactor approaches its nominal power level, the operators partially reinsert the control rods, balancing out the neutron population over time. At this point the reactor is maintained in a critical state, or what is known as steady-state operation. When a reactor is to be shut down, operators fully insert the control rods, inhibiting fission from occurring and forcing the reactor to go into a subcritical state.

Reactor Control

A commonly used parameter in the nuclear industry is reactivity, which is a measure of the state of a reactor in relation to where it would be if it were in a critical state. Reactivity is positive when a reactor is supercritical, zero at criticality, and negative when the reactor is subcritical. Reactivity may be controlled in various ways: by adding or removing fuel, by altering the ratio of neutrons that leak out of the system to those that are kept in the system, or by changing the amount of absorber that competes with the fuel for neutrons. In the latter method the neutron population in the reactor is controlled by varying the absorbers, which are commonly in the form of movable control rods (though in a less commonly used design, operators can change the concentration of absorber in the reactor coolant). Changes of neutron leakage, on the other hand, are often automatic. For example, an increase of power will cause a reactor's coolant to reduce in density and possibly boil.

This decrease in coolant density will increase neutron leakage out of the system and thus reduce reactivity—a process known as negative-reactivity feedback. Neutron leakage and other mechanisms of negative-reactivity feedback are vital aspects of safe reactor design.

A typical fission interaction takes place on the order of one picosecond (10^{-12} second). This extremely fast rate does not allow enough time for a reactor operator to observe the system's state and respond appropriately. Fortunately, reactor control is aided by the presence of so-called delayed neutrons, which are neutrons emitted by fission products some time after fission has occurred. The concentration of delayed neutrons at any one time (more commonly referred to as the effective delayed neutron fraction) is less than 1 percent of all neutrons in the reactor. However, even this small percentage is sufficient to facilitate the monitoring and control of changes in the system and to regulate an operating reactor safely.

Fissile and Fertile Materials

All heavy nuclides have the ability to fission when in an excited state, but only a few fission readily and consistently when struck by slow (low-energy) neutrons. Such species of atoms are called fissile. The most prominently utilized fissile nuclides in the nuclear industry are uranium^{-233} (^{233}U), uranium^{-235} (^{235}U), plutonium^{-239} (^{239}Pu), and plutonium^{-241} (^{241}Pu). Of these, only uranium^{-235} occurs in a usable amount in nature—though its presence in natural uranium is only some 0.7204 percent by weight, necessitating a lengthy and expensive enrichment process to generate a usable reactor fuel.

As an alternative to processing and enriching uranium-235, it is possible to go through the process of generating quantities of other fissile nuclides that are not as prevalent as uranium-235. Prominent sources of these nuclides are thorium-232 (^{232}Th), uranium-238 (^{238}U), and plutonium-240 (^{240}Pu), which are known as fertile materials owing to their ability to transform into fissile materials. For example, thorium-232, the predominant isotope of natural thorium, can be used to generate uranium-233 through a process known as neutron capture. When a nucleus of thorium-232 absorbs, or "captures," a neutron, it becomes thorium-233, whose half-life is approximately 21.83 minutes. After that time the nuclide decays through electron emission to protactinium-233, whose half-life is 26.967 days. The protactinium-233 nuclide in turn decays through electron emission to yield uranium-233.

Neutron capture may also be used to create quantities of plutonium-239 from uranium-238, the principal constituent of naturally occurring uranium. Absorption of a neutron in the uranium-238 nucleus yields uranium-239, which decays after 23.47 minutes through electron emission into neptunium-239 and ultimately, after 2.356 days, into plutonium-239.

If desired, plutonium-241 may be generated directly through neutron capture in plutonium-240, following the formula $^{240}\text{Pu} + {}^{1}\text{n} = {}^{241}\text{Pu}$.

A power reactor contains both fissile and fertile materials. The fertile materials partially replace fissile materials that are destroyed by fission, thus permitting the reactor to run longer before the amount of fissile material decreases to the point where criticality is no longer manageable. Plutonium-240 is particularly found to build up in reactors after long periods of operation, as it has a longer half-life than all its parent nuclides.

Heat Removal

The Angra nuclear power plant, using pressurized-water reactors, at Angra dos Reis, near Rio de Janeiro, Brazil.

A significant portion of the energy of fission is converted to heat the instant that the fission reaction breaks the initial target nucleus into fission fragments. The bulk of this energy is deposited in the fuel, and a coolant is required to remove the heat in order to maintain a balanced system (and also to transfer the heat energy to the power-generating plant). The most common coolant is water, though any fluid can be used. Heavy water (deuterium oxide), air, carbon dioxide, helium, liquid sodium, sodium-potassium alloy (called NaK), molten salts, and hydrocarbons have all been used in reactors or reactor experiments.

Some research reactors are operated at very low power and have no need for a dedicated cooling system; in such units the small amount of generated heat is removed by conduction and convection to the environment. Very high power reactors, on the other hand, must have extremely sophisticated cooling systems to remove heat quickly and reliably; otherwise, the heat will build up in the reactor fuel and melt it. Indeed, most reactors operate on the principle that their fuel cannot be allowed to melt; therefore, the systems designed to cool the fuel must operate sufficiently under both normal and abnormal conditions. Systems that enable sufficient cooling during all credible abnormal conditions in nuclear power plants are referred to as emergency core-cooling systems.

Shielding

An operating reactor is a powerful source of radiation, since fission and subsequent radioactive decay produce neutrons and gamma rays, both of which are highly penetrating radiations. A reactor must have specifically designed shielding around it to absorb and reflect this radiation in order to protect technicians and other reactor personnel from exposure. In a popular class of research reactors known as "swimming pools," this shielding is provided by placing the reactor in a large, deep pool of water. In other kinds of reactors, the shield consists of a thick concrete structure around the reactor system referred to as the biological shield. The shield also may contain heavy metals, such as lead or steel, for more effective absorption of gamma rays, and heavy aggregates may be used in the concrete itself for the same purpose.

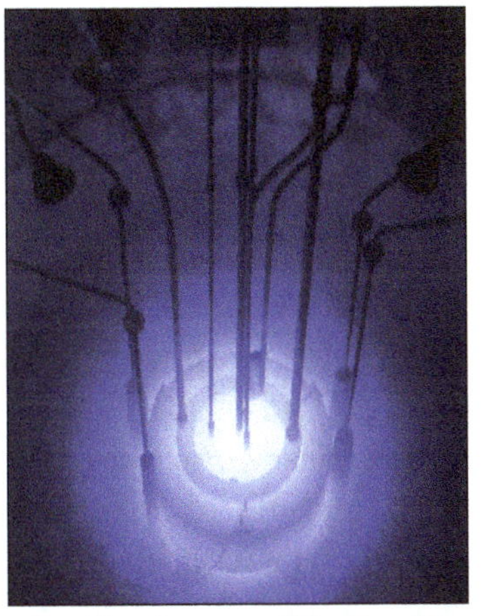

Cherenkov radiation emitted by the core of the Reed Research Reactor located at Reed College in Portland, Oregon, U.S.

Critical Concentration and Size

Not every arrangement of material containing fissile fuel can be brought to criticality. Even if a reactor was designed such that no neutrons could leak out, a critical concentration of fissile material would have to be present in order to bring the reactor to a critical state. Otherwise, absorption of neutrons by other constituents of the reactor might dominate and inhibit a sustained chain fission reaction. Similarly, even where there is a high-enough concentration for criticality, the reactor must occupy an appropriate volume and be of a prescribed geometric form, or else more neutrons will leak out than are created through fission. This requirement imposes a limit on the minimum critical volume and critical mass within a reactor.

Although the only useful fissile material in nature, uranium-235, is found in natural uranium, there are only a few combinations and arrangements of this and other materials that enable a reactor to maintain a critical state for a period of time. To increase the range of feasible reactor designs, enriched uranium is often used. Most of today's power reactors employ enriched uranium fuel in which the percentage of uranium-235 has been increased to between 3 and 5 percent, approximately five and a half times the concentration in natural uranium. Large plants for enriching uranium exist in several countries; indeed, enrichment has become a commercial enterprise.

Thermal, Intermediate and Fast Reactors

Reactors are conveniently classified according to the typical energies of the neutrons that cause fission. Neutrons emanating in fission are very energetic; their average energy is around two million electron volts (MeV), nearly 80 million times the energy of atoms in ordinary matter at room temperature. As neutrons scatter or collide with nuclei in a reactor, they lose energy. This action is referred to as down-scattering. The choice of reactor materials and of fissile material concentrations determines the rate at which neutrons are slowed through down-scattering before causing fission.

In a thermal reactor, most neutrons down-scatter in the moderator material before interacting with a fissile material. Down-scattering events take place until the neutrons have reached thermal equilibrium with the reactor at energies of a few hundredths of an electron volt. Neutrons lose energy most efficiently by colliding with light atoms such as hydrogen (mass 1), deuterium (mass 2), beryllium (mass 9), and carbon (mass 12). For this reason, materials that contain atoms of these elements—water, heavy water, beryllium metal and oxide, and graphite—are deliberately incorporated into a thermal reactor and are known as moderators. Since water and heavy water also can function as coolants, they perform a dual purpose in thermal reactors.

One disadvantage of thermal reactors is that at low energies uranium-235 and plutonium-239 not only can be fissioned by thermal (or slow) neutrons but also can capture neutrons without undergoing fission. Neutron capture transforms these nuclides into, respectively, uranium-236 and plutonium-240, which are not fissile. The probability of neutron capture is much lower at higher energy levels than at thermal energies. To achieve higher energy levels and promote fission over neutron capture, a reactor can be built to operate without a moderator. Depending on the number of scattering events that take place with heavier atoms before fission occurs, the typical fission-causing neutrons may have energies in the range of 0.5 electron volt to thousands of electron volts (intermediate reactors) or several hundred thousand electron volts (fast reactors). Such reactors require higher concentrations of fissile material to reach criticality than do reactor designs that operate at thermal energy levels; however, they are more efficient at converting fertile material to fissile material. Fast reactors can be designed to produce more than one new fissile atom for each fissile atom destroyed. Such reactors are referred to as breeder reactors. Breeder reactors may become important if world demand for nuclear power turns out to be long-term and if access to naturally available sources of fissile material becomes limited.

Reactor Design and Components

There are a large number of ways in which a nuclear reactor may be designed and constructed; many types have been experimentally realized. Over the years, nuclear engineers have developed reactors with solid and liquid fuels, thick- and no-reflectors, forced cooling circuits and natural conduction or convection heat-removal systems, and so on. Most reactors, however, have certain basic components.

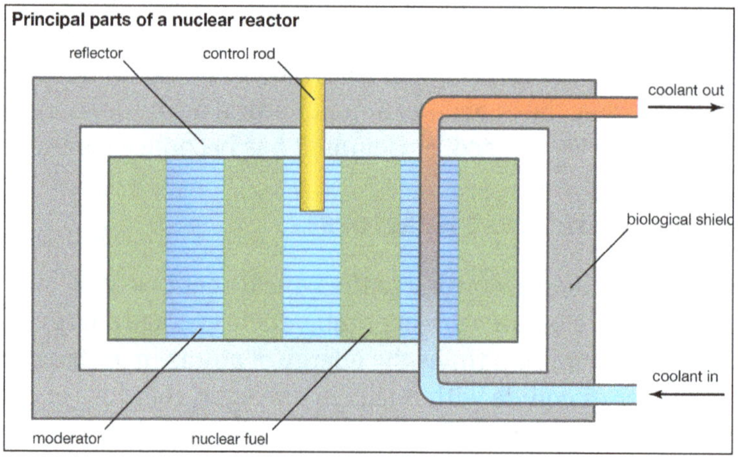

The basic components of a nuclear reactor.

Core

All reactors have a core, a central region that contains the fuel, fuel cladding, coolant, and (where separate from the latter) moderator. The fission energy in a nuclear reactor is produced in the core.

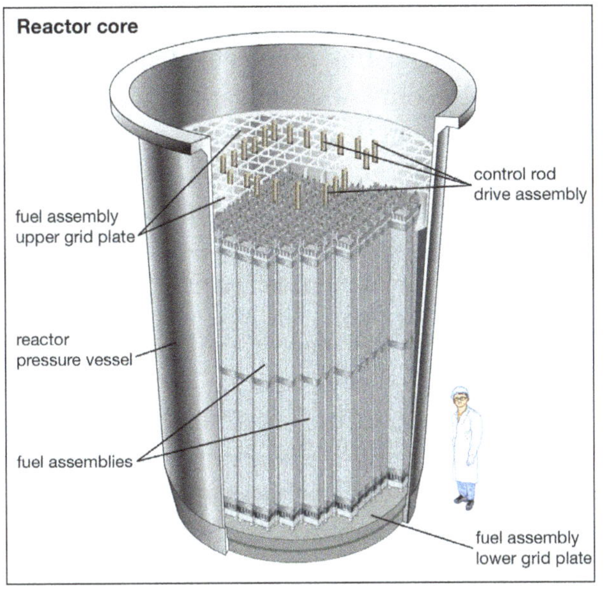

The core of a pressurized-water nuclear reactor.

The fuel is usually heterogeneous—i.e., it consists of elements containing fissile material along with a diluent. This diluting agent may be fertile material or simply material that has good mechanical and chemical properties and does not readily absorb neutrons. All diluents act as a matrix in which the fissile material can stably reside through its operable life. In solid fuels, the diluted fissile material is enclosed in a cladding—a substance that isolates the fuel from the coolant and minimizes the likelihood that radioactive fission products will be released. Cladding is often referred to as a reactor's first fission product barrier, as it is the first barrier that fissile material contacts after nuclear fission.

Fuel Types

A reactor's fuel must conform to the integral design of the reactor as well as the mechanisms that drive its operations.

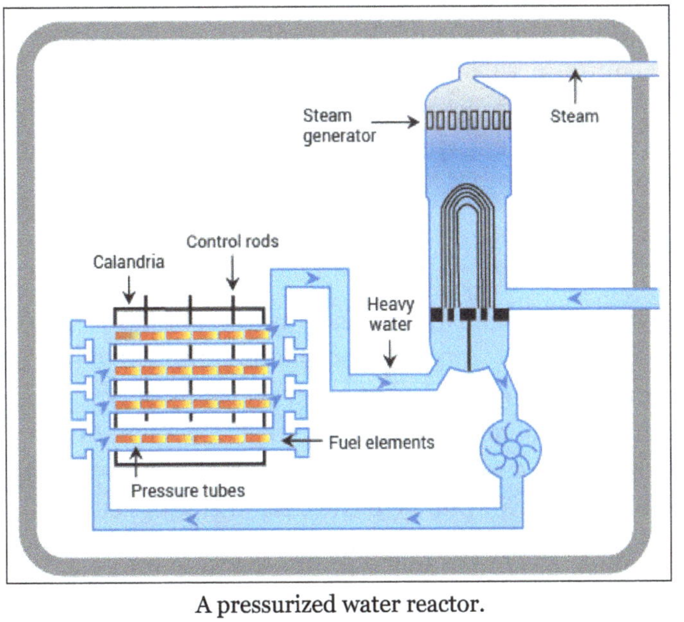

A pressurized water reactor.

The light-water reactor (LWR), which is the most widely used variety for commercial power generation in the world, employs a fuel consisting of pellets of sintered uranium dioxide loaded into cladding tubes of zirconium alloy or some other advanced cladding material. The tubes, called pins or rods, measure approximately 1 cm (less than half an inch) in diameter and roughly 3 to 4 metres (10 to 13 feet) in length. The tubes are bundled together into a fuel assembly, with the pins arranged in a square lattice. The uranium used in the fuel is 3 to 5 percent enriched. Since light (ordinary) water, used in LWRs as both the coolant and the moderator, tends to absorb more neutrons than other moderators do, such enrichment is crucial.

The CANDU (Canada Deuterium Uranium) reactor, which is the principal type of heavy-water reactor, uses natural uranium compacted into pellets. These pellets are inserted in long tubes and arranged in a lattice. A CANDU reactor fuel assembly measures approximately 1 metre (almost 40 inches) in length. Several assemblies are arranged end-to-end within a channel inside the reactor core. The use of heavy water rather than light water as the moderator enhances the scattering of neutrons rather than their capture, thereby increasing the probability of fission with the fuel material.

In one version of the high-temperature graphite reactor, the fuel is constructed of small spherical particles, or microspheres, containing uranium dioxide at the centre with concentric shells of carbon, silicon carbide, and carbon around them. These shells serve as localized cladding for each fuel sphere. The particles are then mixed with graphite and encased in a macroscopic graphite cladding.

In a sodium-cooled fast reactor, commonly called a liquid-metal reactor (LMR), the fuel consists of uranium dioxide or uranium-plutonium dioxide pellets (French design) or of uranium-plutonium-zirconium metal alloy pins (U.S. design) in steel cladding.

The most common type of fuel used in research reactors consists of plates of a uranium-aluminum alloy with an aluminum cladding. The uranium is enriched to slightly less than 20 percent, while silicon and aluminum are included in the "meat" of the plate to serve as the diluent and fuel matrix. Although aluminum has a lower melting point than other cladding materials, the flat plate design maintains a low fuel temperature, as the plates are often barely more than 1.25 mm thick. A common variety of research reactor known as TRIGA (from training, research, and isotope-production reactors–General Atomic) employs a fuel of mixed uranium and zirconium hydride, often doped with small concentrations of erbium and the whole clad in stainless steel.

Coolants and Moderators

A variety of substances, including light water, heavy water, air, carbon dioxide, helium, liquid sodium, liquid sodium-potassium alloy, and hydrocarbons (oils), have been used as coolants. Such substances are, in general, good conductors of heat, and they serve to carry the thermal energy produced by fission from the fuel and through the integral system, finally either venting the heat directly to the atmosphere (in the case of research reactors) or transporting it to the steam-generating equipment of the nuclear power plant (in the case of power reactors).

In many cases, the same substance functions as both coolant and moderator, as in the case of light and heavy water. The moderator slows the fast (high-energy) neutrons emitted during fission to energies at which they are more likely to induce fission. In doing so, the moderator helps initiate and sustain a fission chain reaction.

Reflectors

A reflector is a region of unfueled material surrounding the core. Its function is to scatter neutrons that leak from the core, thereby returning some of them back into the core. This design feature allows for a smaller core size. In addition, reflectors "smooth out" the power density by utilizing neutrons that would otherwise leak out through fissioning within fuel material located near the core's outer region.

The reflector is particularly important in research reactors, since it is the region in which much of the experimental apparatus is located. In some research reactor designs, reflectors are located inside the core as central islands in which high neutron intensities can be achieved for experimental purposes.

In most types of power reactors, a reflector is less important; this is due to the reactor's large size, which reduces the proportion of neutrons that may leak from the core region. The liquid-metal reactor represents a special case. Most sodium-cooled reactors are deliberately built to allow a large fraction of their neutrons—those not needed to maintain the chain reaction—to leak from the core. These neutrons are valuable because they can produce new fissile material if they are absorbed by fertile material. Thus, fertile material—generally depleted uranium or its dioxide—is placed around the core to catch the leaking neutrons. Such an absorbing reflector is referred to as a blanket or a breeding blanket.

Reactor Control Elements

All reactors need unique elements for control. Although control can be achieved by varying parameters within the coolant circuit or by varying the amount of absorber dissolved in the coolant or moderator, by far the most common method utilizes absorbing assemblies—namely, control rods or, in some cases, blades. Typically a reactor is equipped with three types of rods for different purposes: (1) safety rods for starting up and shutting down the reactor, (2) regulating rods for adjusting the reactor's power rate, and (3) shim rods for compensating for changes in reactivity as fuel is depleted by fission and neutron capture.

The most important function of the safety rods is to shut down the reactor, either when such a shutdown is scheduled or in case of a real or suspected emergency. These rods contain enough absorber to terminate a chain reaction under any conceivable condition. They are withdrawn before fuel is loaded and remain available in case a loading error requires their action. After the fuel is loaded, the rods are inserted, to be withdrawn again when the reactor is ready for operation. The mechanism by which they are moved is designed to be fail-safe in the sense that if there is a mechanical failure, the safety rods will fall by gravity or some other safe means into the core. In some cases, moreover, the safety rods have an automatic feature, such as a fuse, which releases them by virtue of physical effects independent of electronic signals.

Regulating rods are deliberately designed to affect reactivity only by a small degree. It is assumed that at some time the rods might be totally withdrawn by mistake, and the idea is to keep the added reactivity in such cases well within sensible limits. A well-designed regulating rod will add so little reactivity when it is removed that the delayed neutrons will continue to control the rate of power increase.

Shim rods are designed to compensate for the effects of burnup (i.e., energy production). Reactivity changes resulting from burnup can be large, but they occur slowly over periods of days to years, as compared with the seconds-to-minutes range over which safety actions and routine regulation take place. Therefore, shim rods may control a significant amount of reactivity, but they will work optimally only when constraints are imposed on their speed of movement. A common way in which shims are operated is by inserting or removing them as regulating rods reach the end of their most useful position range. When this happens, shim rods are moved so that the regulating rods can be reset.

The functions of shim and safety rods are sometimes combined in rods that have low rates of withdrawal but that can be rapidly inserted. This is usually done when the effect of burnup decreases reactivity. The rods are only partially inserted at the outset of operation. However, in the event that the system must be shut down as quickly as possible, the reactor operator may "scram" the reactor, fully dropping the control rods into the core and immediately sending the reactor into a subcritical state. (The expression "scram" is said to stand for "safety control rod axe man," a reference to ad-hoc emergency preparations made for the earliest nuclear reactor.)

The amount of shim control required can be reduced by the use of a burnable "poison." This is a neutron-absorbing material, such as boron or gadolinium, that burns off faster than the fissile material does over the lifetime of the core. At the beginning of operation, the inclusion of a burnable poison regulates the extra reactivity that has been built into the fuel to compensate for the amount of fuel consumed. At the end of an operating period, the absorbing material is often completely transformed through neutron capture.

Some boiling-water reactors utilize cruciform (T-shaped) control blades as the neutron-absorbing control mechanism. Because a number of these reactor vessels are designed with internal components above the core region, the control blades are inserted from below the core. Control blades operate on the same principle as control rods. However, since they are inserted upward into the core, they cannot use gravity to fall into place and put the reactor into a subcritical state in the event of a loss of power or some other abnormal condition. For this reason, control blades are connected to hydraulic drives that force compressed air into the mechanism upon initiation, injecting the control blades into the core.

Structural Components

The structural components of a reactor hold the system together and permit it to function as a useful energy source. The most important structural component in a nuclear power plant is usually the reactor vessel. In both the light-water reactor and the high-temperature gas-controlled reactor (HTGR), a reactor pressure vessel (RPV) is utilized so that the coolant is contained and operated under conditions appropriate for power generation—namely, elevated temperature and pressure. Within the reactor vessel are a number of structural elements: grids for holding the reactor core and solid reflectors, control-rod guide tubes, internal thermal hydraulic components (e.g., pumps or steam circulators) in some cases, instrument tubes, and components of safety systems.

Coolant System

The function of a power reactor installation is to extract as much heat of nuclear fission as possible and convert it to useful power, generally electricity. The coolant system plays a pivotal role in performing this function. A coolant fluid enters the core at low temperature and exits at a higher temperature after collecting the fission energy. This higher-temperature fluid is then directed to conventional thermodynamic components where the heat is converted into electric power. In most light-water, heavy-water, and gas-cooled power reactors, the coolant is maintained at high pressure. Sodium and organic coolants operate at atmospheric pressure.

Research reactors have very simple heat-removal systems, as their primary purpose is to perform research and not generate power. In research reactors, coolant is run through the reactor, and the heat that is removed is transferred to ambient air or to water without going through a power cycle. In research reactors of the lowest power, running at only a few kilowatts, this may involve simple heat exchange to tap water or to a pool of water cooled by ambient air. During operation at higher power levels, the heat is usually removed by means of a small natural-draft cooling tower.

Containment System

Reactors are designed with the expectation that they will operate safely without releasing radioactivity to their surroundings. It is, however, recognized that accidents can occur. An approach using multiple fission product barriers has been adopted to deal with such accidents. These barriers are, successively, the fuel cladding, the reactor vessel, and the shielding. As a final barrier, the reactor is housed in a containment structure, often simply referred to as the containment.

Installation of the dome of a containment structure at the Taishan nuclear power plant, Guangdong province, southern China, 2012.

Containment Design Principles

The containment basically consists of the reactor building, which is designed and tested to prevent elevated levels of radioactivity that might be released from the fuel cladding, the reactor vessel, and the shielding from escaping to the environment. To meet this purpose, the containment structure must be at least nominally airtight. In practice, it must be able to maintain its integrity under circumstances of a drastic nature, such as accidents in which most of the contents of the reactor core are released to the building. It has to withstand pressure buildups and damage from debris propelled by an energy burst within the reactor, and it must pass appropriate tests to demonstrate that it will not leak more than a small fraction of its contents over a period of several days, even when its internal pressure is well above that of the surrounding air. The containment building also must protect components located inside it from external forces such as tsunamis, tornadoes, and airplane crashes.

The most common form of containment building is a cylindrical structure with a spherical dome, which is characteristic of LWR systems. This structure is much more typical of nuclear plants than the large cooling tower that is often used as a symbol for nuclear power. (It should be noted that cooling towers are found at large modern coal- and oil-fired power stations as well.)

Reactors other than those of the LWR type also have containment structures, though they vary in shape and construction. When it can be justified that major pressure buildups are not to be expected, the containment may be any functional form of airtight structure. In the United States and a number of other countries around the world, containment structures are required for all commercial power reactors and all high-power research reactors. In general, low-power research reactors are exempt, on the basis of the common assumption that an accident in such systems will not lead to a widespread release of radioactivity. In the United States, reactors operated by the Department of Energy and by the armed services are also exempt, a matter that has caused considerable controversy. Some of these have containment structures, whereas others do not.

Containment Systems and Major Nuclear Accidents

The concept of the containment originated in the United States during the 1950s and was generally accepted throughout much of the world. The Soviet-bloc countries, however, did not concur with this view, and when containment was added to Soviet reactor designs, it was generally not up to Western standards. For example, during the Chernobyl accident of 1986 in Ukraine, the power station's Unit 4, which suffered a catastrophic explosive accident and fire, had an internal structure that could withstand the loss of function of only a single pressure tube. Though the structure was called a containment, this was a misnomer by Western standards, and the structure would more suitably be referred to as a confinement.

Severe tests of Western-style containment systems occurred during the Three Mile Island accident in the United States in 1979 and the Fukushima accident in Japan in 2011. At Three Mile Island Unit 2, near Harrisburg, Pennsylvania, a stoppage of core cooling resulted in the destruction, including partial melting, of the entire core and the release of a large part of its radioactivity to the enclosure around the reactor—that is, the containment. In spite of a hydrogen deflagration that also occurred during the accident, the containment structure prevented all but a very small amount of radioactivity from entering the environment and is credited with having prevented a major radioactive release and its consequences.

At the Fukushima Daiichi ("Number One") plant in northeastern Honshu, Japan, a loss of main and backup power after an earthquake and tsunami led to a partial meltdown of fuel rods in three reactors. Melted material bored small holes in the lower head of two reactor pressure vessels;

one of these was punctured again by an explosion. Radioactive water was soon discovered to have leaked into the ocean through cracks in the foundation of the containment. Within a few weeks, the cracks had been sealed, and within six months the reactors had been stabilized to the point where workers could begin to enter the containment. Despite a catastrophic sequence of natural events that led to the accident, the fission product barriers served their design purpose and kept all but a small amount of fission products from entering the environment.

Types of Reactors

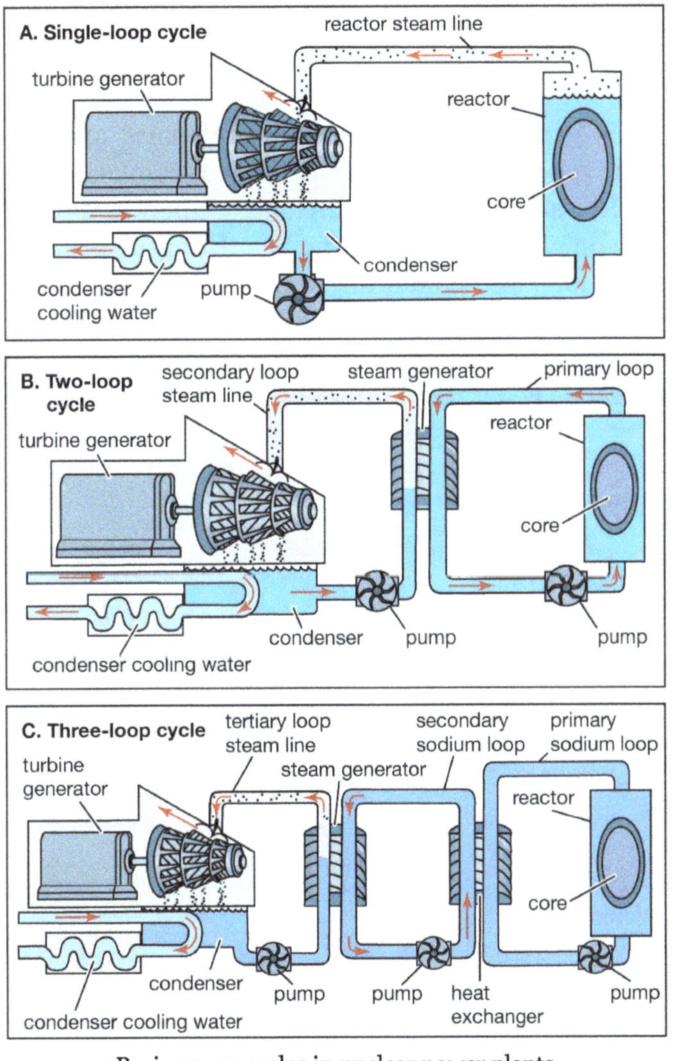

Basic power cycles in nuclear power plants.

In figure, (A) Single-loop cycle; as shown, it represents a boiling-water reactor (BWR), but it could also represent a direct-cycle, high-temperature gas-cooled reactor (HTGR) if helium were substituted for steam. (B) Two-loop cycle; the primary loop depicted here could constitute a pressurized-water reactor (PWR), a CANDU pressurized heavy-water reactor (PHWR), or a helium HTGR. (C) Three-loop cycle; this is found only in sodium-cooled reactors where an intermediate loop of nonradioactive sodium is provided between the radioactive primary loop and the steam generator.

Most of the world's existing reactors are power reactors, providing the heat needed to turn turbines that run electric-power generators. There are also numerous research reactors, and some navies of the world have submarines or surface ships driven by propulsion reactors. There are several types of power reactors, but only one, the light-water reactor, is widely used. Other significant types are briefly described, as are research and propulsion reactors. Some attention is also given to the prospective uses of reactors for space travel and for certain industrial purposes.

Power Reactors

Light-water Reactors

1. PWRs and BWRs:

Light-water reactors (LWRs) are power reactors that are cooled and moderated with ordinary water. There are two basic types: the pressurized-water reactor (PWR) and the boiling-water reactor (BWR). In the PWR, water at high pressure and temperature removes heat from the core and is transported to a steam generator. There the heat from the primary loop is transferred to a lower-pressure secondary loop also containing water. The water in the secondary loop enters the steam generator at a pressure and temperature slightly below that required to initiate boiling. Upon absorbing heat from the primary loop, however, it becomes saturated and ultimately slightly superheated. The steam thus generated ultimately serves as the working fluid in a steam-turbine cycle.

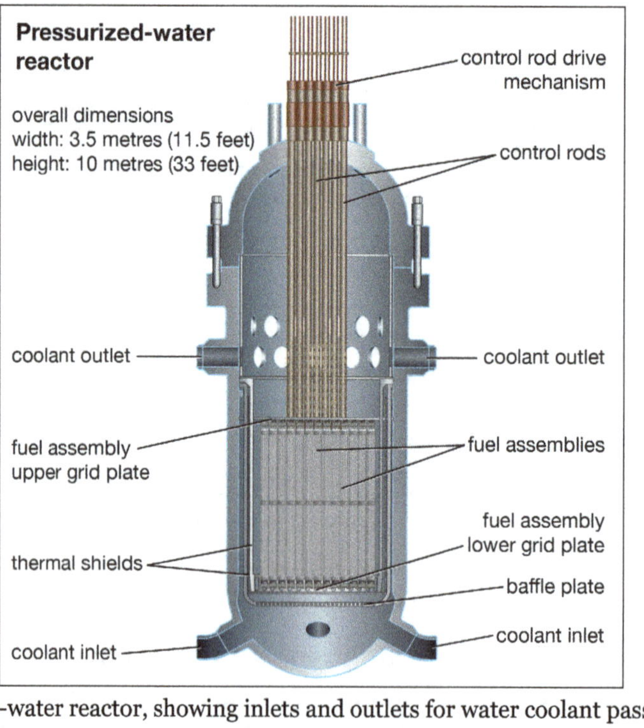

Section of a pressurized-water reactor, showing inlets and outlets for water coolant passing through the core.

A BWR operates on the principle of a direct power cycle. Water passing through the core is allowed to boil at an intermediate pressure level. The saturated steam that exits the core region is transported through a series of separators and dryers located within the reactor vessel that promote a superheated state. The superheated water vapour is then used as the working fluid to turn the steam turbine.

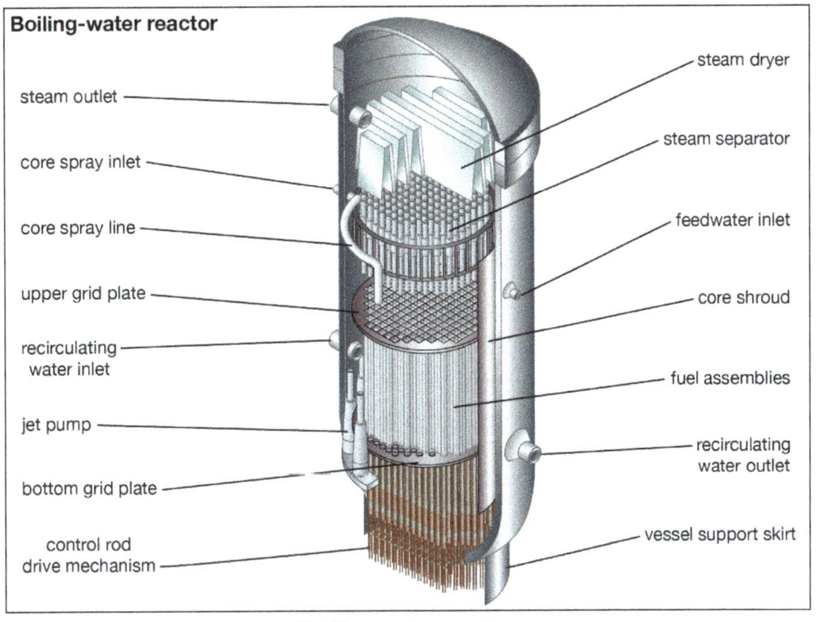

Boiling-water reactor.

Advantages and Disadvantages of PWRs and BWRs

Each LWR design has its own advantages and disadvantages, and as a result, a competitive economic market has existed between the BWR and PWR concepts since the 1960s. For instance, although there are fewer mechanical components in the steam cycle of a BWR design, additional components are required to support the reactor's emergency core-cooling system. Furthermore, the BWR vessel's internal system is more complex, since it includes internal recirculation pumps and complex steam separation and drying equipment that are not found in a PWR design. On the other hand, even though the internals of the PWR are simpler, a BWR power plant is smaller, because it has no steam generators. In fact, the steam generators of a PWR—there are typically four of them in a big plant—are larger than the reactor vessel itself.

The direct-cycle philosophy of a BWR design reduces heat loss between the core and the steam turbine, but the BWR operates at lower pressures and temperatures than the PWR, giving it less thermodynamic efficiency. Furthermore, because the BWR's power density is somewhat lower than that of the PWR, the pressure vessel must be built to a larger diameter for the same reactor power. On the other hand, because the BWR operates at lower pressure, its pressure vessel is thinner than the pressure vessel of a PWR.

2. Fueling and Refueling LWRs:

Both standard light-water designs are fueled with uranium dioxide pellets in zirconium alloy cladding. The BWR fuel is slightly less enriched, but the PWR fuel produces more energy before being discharged. The control rods of a typical PWR are inserted from the top (through the reactor head), whereas those of a BWR are inserted from the bottom.

Light-water reactors are refueled by removing the reactor head—after lowering and unlatching the safety rods in the case of a PWR. This exposes the reactor to visual observation. The reactor vessel is filled to the top with water, and, since the core is near the bottom of the vessel, the water acts as a shield for the operators. Each fuel assembly to be removed is grasped by a fuel-handling tool and then lifted from its position in the core into a shielded cask, within which the assembly is transferred to a storage pool for cooling while it is still highly radioactive. This process is repeated for each element that requires removal.

In most LWRs a typical refueling cycle removes approximately one-third of the fuel assemblies. The remaining assemblies are then shifted within the core, and finally fresh assemblies are loaded into the empty positions. The purpose of shifting fuel at the time of reload is to achieve an optimal reactivity and power distribution for the next cycle of operation. Reloading is a time-consuming operation. In principle it could be accomplished in two weeks, but in practice nuclear power plants undergo maintenance during reload, which often takes considerably more time—up to several months. Utilities schedule maintenance and reload during the spring and fall, when electricity demand is lowest and the electrical grid usually has reserve capacity.

The removed fuel stored in the storage pool not only is highly radioactive but also continues to produce energy. This energy is removed by natural circulation of the water in the storage pool. During the 1960s, when the nuclear industry was in its early stage, it was expected that spent fuel could be shipped out for reprocessing within two years. However, this option is currently practiced only in a few countries around the world—notably France, the United Kingdom, and Japan, where large-scale facilities employ a well-developed reprocessing technique known as PUREX. Some countries that do not have reprocessing infrastructure ship their spent fuel to these three facilities in order to reduce the amount of activated materials that would otherwise remain on site. In the United States, which does not reprocess, storage pools have continued to receive spent fuel, and some of the pools are filling up. Options available to nuclear plant operators are to store the spent fuel more densely than originally planned, to build new pools, or to store the oldest in aboveground silos (dry storage) locally on site. Ultimately, this fuel will be transferred to the U.S. Department of Energy for reprocessing or waste disposal or both, though a viable disposal program has not yet been established.

3. Global Status of LWRs:

During the 1970s, light-water reactors represented the cheapest source of new electricity in most parts of the world, and it still is economical in many countries such as Japan, South Korea, Taiwan, France, and China (which, in the 1990s, embarked on an ambitious program of building nuclear power plants, almost all of them using light-water technology). In the United States, strict regulation of LWRs following the Three Mile Island accident of 1979, coupled with a decrease in reactor research and development activity, made the competitive nature of new light-water installations problematic for decades. In the early 21st century, signs began to emerge of a reviving nuclear power sector in the United States as demand for reliable energy sources continued to increase and it became evident that the fleet of U.S. nuclear power plants was aging.

Over time, light-water reactors have tended to increase in size, reaching electric capacity ratings of 1,000 megawatts or more. More recent designs—small modular reactors that produce 300 megawatts or less—may be able to provide energy on a more economic scale and thus appeal to a much broader market.

High-temperature Gas-cooled Reactors

The high-temperature gas-cooled reactor (HTGR), as mentioned above in Fuel types, is fueled by a mixture of graphite and fuel-bearing microspheres. There are two competitive designs of this reactor type: (1) a German "pebble bed" system that uses spherical fuel elements, nominally 60 mm (2.5 inches) in diameter, containing a graphite-and-fuel mixture coated in a graphite shell; and (2) an American version in which the fuel is loaded into precisely located graphite hexagonal prisms that interlock to create the core of the vessel. In both variants, the coolant consists of helium pressurized to approximately 100 bars, or roughly 100 standard atmospheres. In the German system, the helium passes through interstices in the bed of the spherical fuel elements, whereas in the American system, it passes through holes in the graphite prisms that align along the axis of the core region in the reactor vessel. Both are capable of operating at very high temperature, since graphite has an extremely high sublimation temperature and helium is completely inert chemically. The hot helium can be used directly as the working fluid in a high-temperature gas turbine, or its heat can be utilized to generate steam for a water cycle.

Experimental prototypes of both the American and German designs were built and operated from the 1960s to the 1980s, but by the early 1990s, they had resulted in no orders for commercial plants. In 2000 the High Temperature Reactor-10 (HTR-10), a Chinese reactor based on the German pebble-bed design, began operating at a nominal steady-state power level of 10 megawatts. Intended to test the viability of a full-sized HTGR power plant, the HTR-10 provided power and heating for the campus of Tsinghua University near Beijing.

Liquid-metal Reactors

Sodium-cooled fast-neutron-spectrum liquid-metal reactors (LMRs) received much attention during the 1960s and '70s when it appeared that their breeding capabilities would soon be needed to supply fissile material to a rapidly expanding nuclear industry. When it became clear in the 1980s that this was not a realistic expectation, enthusiasm waned. The developmental work of the previous decades, however, resulted in the construction of a number of LMRs around the world—in the United States, Russia, France, Britain, Japan, and Germany. Most LMRs are fueled with uranium dioxide or mixed uranium-plutonium dioxides. In the United States, however, the greatest success has been with metal fuels.

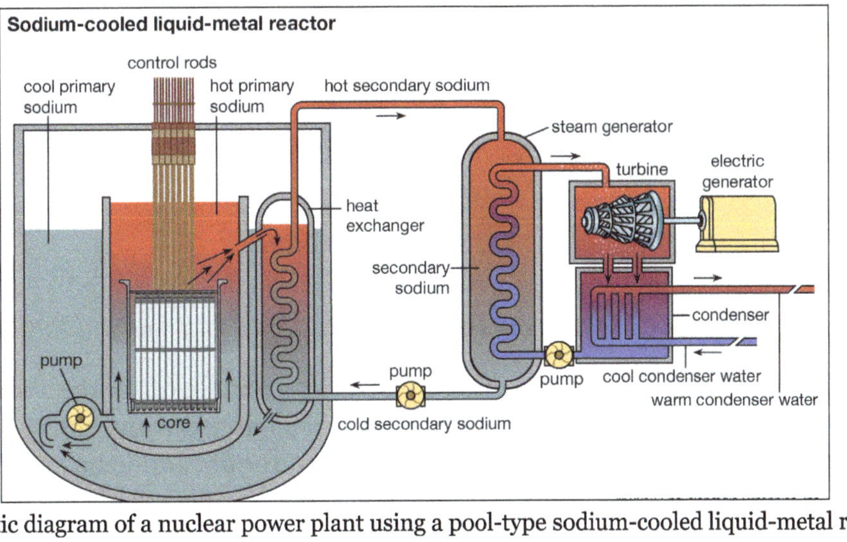

Schematic diagram of a nuclear power plant using a pool-type sodium-cooled liquid-metal reactor.

While some LMRs are of the loop type, equipped with heat exchangers and pumps outside the primary reactor vessel, others are of the pool variety, featuring a large volume of primary sodium in a pool that also contains the primary pumps and the primary-to-secondary heat exchanger. In all types the heat extracted from the core by primary sodium is transferred to a secondary, non-radioactive sodium loop, which serves as the heat source for a steam generator that heats the water in a tertiary loop to power a turbine. The pool type seems to have some safety advantage in that the large volume of primary sodium heats up only slowly even if no power is extracted; thus, the reactor is effectively isolated from upsets in the balance of the plant. The reactor core in all such systems is a tightly packed bundle of fuel in steel cladding through which the sodium coolant flows to extract the heat. Most LMRs are breeders or are capable of breeding, which is to say that they all produce more fissile material than they consume.

CANDU Reactors

Canada has focused its developmental efforts on reactors that utilize abundant domestic natural uranium as fuel without having to resort to enrichment services that would be supplied only by other countries. The result of this policy is the Canada Deuterium Uranium (CANDU) reactor—a line of natural uranium-fueled reactors moderated and cooled by heavy water. A reactor of this kind consists of a tank, or calandria vessel, containing a cold heavy water moderator at atmospheric pressure. The calandria is pierced by pressure tubes made of zirconium alloy in which the natural uranium fuel is placed and through which heavy water coolant is circulated. In contrast to the more common LWR reactor vessel, which is aligned in a vertical configuration, a CANDU reactor vessel is aligned horizontally. Electric-power generation is obtained through the transfer of the heat of fission into the heavy-water coolant, which is circulated to a steam generator. The light-water steam exiting the steam generator in the secondary loop is then transported through a conventional turbine cycle.

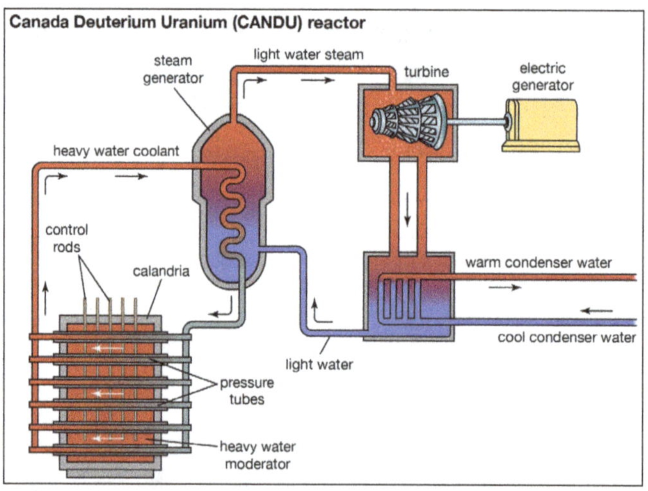

Schematic diagram of a nuclear power plant using a Canada Deuterium Uranium (CANDU) reactor.

The fuel assembly of a CANDU reactor, which consists of a bundle of short zirconium alloy-clad tubes containing natural uranium dioxide pellets, can be changed while the system is running. A new assembly is simply pushed into one end of a pressure tube, and the spent fuel that it replaces is collected as it is extruded at the opposite end. This feature has given the CANDU higher capacity factors than other reactor types. Capacity factor refers to the ratio of time that a reactor is operating at full power during a given period versus the total available time during that same period. Thus, a capacity factor of 1.0 refers to a reactor that operates at full power 24 hours a day over the entire period of time being considered. Several countries have purchased CANDU reactors for the same reason that they were developed by Canada—to be independent of foreign enrichment services.

Advanced Gas-cooled Reactor

The advanced gas-cooled reactor (AGR) was developed in the United Kingdom as the successor to reactors of the Calder Hall class, which combined plutonium production and power generation. Calder Hall, the first nuclear station to feed an appreciable amount of power into a civilian network opened in 1956. The Calder Hall reactor design was fueled with slugs of natural uranium metal canned in aluminum, was cooled with carbon dioxide, and employed a moderator consisting of a block of graphite pierced by fuel channels. In the AGR, fuel pins clad in Zircaloy (a trademark for alloys of zirconium having low percentages of chromium, nickel, iron, and tin) and loaded with approximately 2 percent enriched uranium dioxide are placed into zirconium-alloy channels that pierce a graphite moderator block. This design utilizes fast neutron energies and is therefore referred to as a fast reactor. The enriched fuel permits operation to economic levels of fuel burnup. A coolant of carbon dioxide absorbs and transports heat to a steam generator, where the heat is conducted to the secondary loop and ultimately to a traditional steam-turbine cycle.

Dungeness B, a nuclear power plant using an advanced gas-cooled reactor, located at Dungeness Point, Kent, England.

Although a number of AGRs have been built in the United Kingdom, maintenance and malfunction issues have proved to be more costly than expected, and no new AGRs are projected for construction.

Other Power Reactor Types

A large variety of reactor types have been built and operated on an experimental basis. A few examples include organic liquid-cooled and liquid-moderated reactors that can operate like a pressurized-water reactor but without requiring pressures in the primary circuit to be as high as those in the traditional PWR; sodium-cooled graphite-moderated reactors; and heavy-water reactors built in a pressure-vessel design. Each design has its own advantages and disadvantages.

Research Reactors

Water-cooled and Plate-Fuel Reactors

These are the most common type of research reactor. Water-cooled, plate-fuel reactors use enriched uranium fuel in plate assemblies and are cooled and moderated with water. They operate over a wide range of thermal power levels, from a few kilowatts to hundreds of megawatts. As the primary mission of research reactors is not electricity production, they are characterized not by the amount of electric power that they produce but rather by thermal power, neutron density, and nominal neutron energy within the core region. It is these parameters that help quantify a research reactor's ability to perform specific research. The systems with the lowest licensed power levels are usually operated at universities and are used primarily for teaching, whereas those with the highest are used by research laboratories chiefly for materials testing and characterization as well as for general research.

A common form of the water-cooled, plate-fuel reactor is the pool reactor, in which the reactor core is positioned near the bottom of a large, deep pool of water. This has the advantage of simplifying both observation and the placement of channels, commonly referred to as beam ports, from which beams of neutrons can be directed and transported. At lower thermal power levels, no pumping is required, as the natural convection of the coolant past the fuel plates provides sufficient heat removal to maintain a safe operating state. A heat exchanger is usually located at or near the top of the pool, where the hottest water is stratified. At higher operating power levels, pumping becomes necessary to augment the natural circulation.

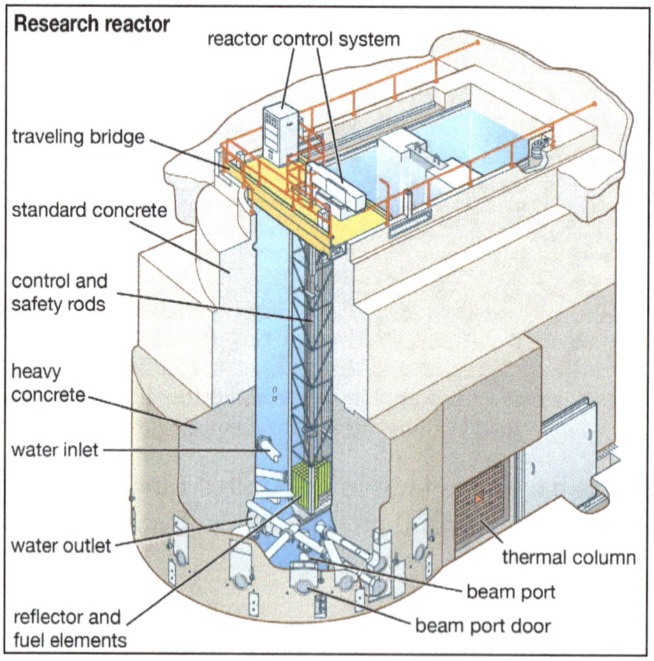

A water-cooled research reactor, with graphite blocks serving as an internal reflector to moderate the reaction.

Most pool reactors use the water of the pool as a reflector, but some have blocks of a solid moderator (canned graphite or beryllium metal) around the core that serves as an inner reflector. The inclusion of graphite and beryllium provides a means for reflecting neutrons back into the core (or a useful location adjacent to the core) that would otherwise leak out of the core. Following interactions between the reflector components, neutrons tend to down-scatter quickly, producing relatively large local regions of thermal neutrons. These regions are often exploited for experimental use by transporting the neutrons down the beam ports and to the experimenter as required.

At higher power levels, it becomes more convenient to employ a tank-type reactor, because it is simpler to control the flow path of pumped water in such a system. Low-power educational reactors also are available in the tank form. The core and reflector arrangement and the position of these components within the tank are similar in both tank-type and pool-type systems. However, solid concrete shielding is increased in robustness around the sides of tank-type reactors, whereas pool-type reactors rely primarily on water as a biological shield.

TRIGA Reactors

The training, research, and isotope-production reactors–General Atomic (TRIGA) system is a popular variety of research reactor. It is another tank-type water-cooled system, but its fuel differs from that employed by the plate-fuel research reactors described above. The fuel element of the TRIGA reactor consists of stainless steel- or aluminum-clad rods containing mixed uranium and zirconium hydrides that are often doped with small concentrations of erbium. In contrast to thin plate-type fuel, TRIGA fuel elements are nominally 3.8 cm (1.5 inches) in diameter and in general approximately 67 cm (26 inches) in total length. A unique characteristic of this fuel is that it exhibits an extremely large negative power-reactivity coefficient—so large that the TRIGA reactor can be placed in an extremely supercritical state for an instant, causing its power to rise very rapidly, after which it quickly shuts itself down on the basis of the fuel's inherent material composition and characteristics. The resulting power transient is referred to as a pulse and is useful for a number of dynamic experiments that require large bursts of neutrons over a short period of time. The total energy released in a pulse does not draw concern toward the safety of the reactor, since the automatic shutdown occurs very quickly and the energy release is proportional to both peak power and pulse duration.

Other Research Reactors

As in the case of power reactors, a number of different reactor types have seen service as research reactors, and a number are still in operation. The variety is so great as to defy cataloging. There have been homogeneous (fueled solution cores), fast, graphite-moderated, heavy-water-moderated, and beryllium-moderated reactors, as well as those adapted to use fuels left over from power reactor experiments. The design of research reactors is much more fluid and sensitive to a greater variety of unique research demands than designs for other applications are.

Ship-propulsion Reactors

The original, and still the major, naval application of nuclear reactors is the propulsion of submarines. The chief advantage of using nuclear reactors for submarine propulsion is that they, unlike fossil-fuel combustion systems, require no air for power generation. Consequently, a nuclear-powered submarine can remain underwater for prolonged periods, whereas a conventional diesel-electric submarine must surface periodically to run its engines in air. Nuclear power confers a strategic advantage on naval surface vessels as well, because it eliminates their dependence on refueling from vulnerable tankers or in foreign ports.

The design of military nuclear power plants is classified for defense security purposes, and so only general information pertaining to them has been published. It is known that U.S. naval power plants are fueled with highly enriched uranium and moderated and cooled with light water. The design of the first nuclear submarine reactor, that of the USS Nautilus, was heavily influenced by high-power research reactor design. Unique features include the incorporation of a very large reactivity margin to accommodate long burnups without refueling and to permit restart after shutdown. For submarine use, the power plant also must be extremely quiet to avoid sonic detection. Various reactor core models have been developed to address the specific requirements of different classes of submarines.

The nuclear-powered submarine USS Nautilus entering New York Harbor, 1958.

The nuclear power plants for U.S. aircraft carriers are believed to have been derived from the power plant designs for the largest submarines, but again the particulars of their design have not been published.

USS Enterprise, commissioned in 1961, the first nuclear-powered aircraft carrier of the U.S. Navy.

Besides the United States, nuclear submarines are deployed by the United Kingdom, France, Russia, China, and India. In each case, the design was developed in secret, but it is generally believed that they are all rather similar; the demands of the application usually lead to similar solutions. Russia also has a small fleet of nuclear-powered icebreakers, whose reactors are thought to be essentially the same as those of the earliest Soviet submarines. As with naval vessels, the ability to operate without refueling is an enormous advantage for Arctic icebreakers.

The Soviet/Russian nuclear-powered icebreaker Lenin, launched in 1957 and in service from 1959 to 1989.

Prototypes of nuclear-powered commercial cargo ships were built and operated by a handful of countries in the latter half of the 20th century, but they were soon decommissioned. These vessels did not operate very economically, and opposition to their docking in a number of major ports was also a factor in their decommissioning. The prototypes were powered by reactors of the pressurized-water type.

Production Reactors

The very first nuclear reactors were built for the express purpose of manufacturing plutonium for nuclear weapons, and the euphemism of calling them production reactors has persisted to this day. At present, most of the material produced by such systems is tritium (3H, or T), the fuel for hydrogen bombs. Plutonium has a long half-life of approximately 24,100 years (specific to plutonium-239), so countries with arsenals of nuclear weapons using plutonium as fissile material generally have more than they expect to need. In contrast, tritium has a half-life of approximately 12 years; thus, stocks of this radioactive hydrogen isotope have to be continuously produced to maintain the required stockpiles. The United States, for example, operates several reactors moderated and cooled by heavy water that produce tritium at the Savannah River facility in South Carolina.

The plutonium isotope that is most desirable for sophisticated nuclear weapons is plutonium-239. If plutonium-239 is left in a reactor for a long time after production, plutonium-240 builds up as an undesirable contaminant. Accordingly, a significant feature of a production reactor is its capability for quick throughput of fuel at a low energy-production level. Any reactor that can be operated this way is a potential production reactor.

The world's first plutonium production reactors, built by the United States in Hanford, Washington, during World War II, were fueled with natural uranium, moderated by graphite, and cooled by light water. It is believed that the early Soviet production reactors were the same sort, and the French and British versions differed only in that they were cooled with gas. The first significant power reactor, the Calder Hall reactor in Cumbria, northwestern England, was actually a dual-purpose production reactor.

Specialized Reactors

Nuclear reactors have been developed to provide electric power and steam heat in far-removed isolated areas. Russia, for instance, operates smaller power reactors specially designed to supply both electricity and steam for heating to accommodate the needs of a number of remote Arctic communities, and in China, the 10-megawatt HTR-10 reactor supplies both heat and electricity for the research institution at which it is located. Independent developmental work on small automatically operated reactors with similar capabilities has been undertaken by Sweden and Canada. Between 1960 and 1972, the U.S. Army used small pressurized-water reactors to provide power for remote bases in Greenland and Antarctica. They were replaced with oil-fired power plants, but it is still technically feasible to employ nuclear power for such applications, as nuclear reactors require less fuel maintenance than a traditional fossil fuel plant and, in general, run at a consistently high capacity. Small modular reactors being designed in the United States might offer unique capabilities such as remote operations.

Reactors have been developed to supply power and propulsion in space. Between 1967 and 1988, the Soviet Union deployed small intermediate reactors in Earth-orbiting satellites (mostly in the Cosmos series) for powering equipment and telemetry, but this policy became a target for criticism. At least one of the Soviet Union's reactor-powered spacecraft reentered the atmosphere and distributed radioactive contamination in remote areas of Canada. The United States launched only one reactor-powered satellite, in 1965, but developmental activity continues for such possible deep-space missions as manned exploration of other planets or the establishment of a permanent lunar base. Reactors for these applications would necessarily be high-temperature systems based on either the HTGR or the LMR design but would use enriched fuel to last the entire life of a prolonged space mission. A power cycle in space must be run at a very high temperature to minimize the size of the radiator from which heat is to be rejected. In addition, a reactor for space applications has to be compact so that it can be shielded with a minimum amount of material and reduce the weight during launch and space flight.

Reactor Safety

Nuclear reactors contain very large amounts of radioactive isotopes—mostly fission products but also such heavy elements as plutonium. If this radioactivity were to escape the reactor, its impact on the people in the vicinity would be severe. The deleterious effects of exposure to high levels of ionizing radiation include increased probability of cancer, cellular damage, an increased number of developmental abnormalities in children exposed in the womb, and even death within a period of several days to months when irradiation is extreme. For this reason, the primary consideration in the design of a reactor is ensuring that a significant release of radioactivity does not occur as a result of any plausible accident scenario. This is accomplished by a combination of preventive measures and mitigating measures. Preventive measures are those that are taken to avoid accidents, and mitigating measures are those that decrease the adverse consequences. In spite of the most stringent preventive and mitigating measures, however, it is still possible that accidents will reach an emergency scale, and in these cases, the nuclear industry and regulators have prepared a set of emergency responses.

Preventive Measures

Design and Operating Standards

Essentially, preventive measures are the set of design and operating rules that are intended to make certain that a reactor is operated safely. The nuclear industry in the United States created a design philosophy referred to as "defense in depth" that numerous other countries have also adopted. In a nuclear power plant following the defense-in-depth model, all safety systems are required to be functionally independent, inherently redundant, and diverse in design.

Among the most well-known preventive measures are the reports and inspections for double-checking that a plant is properly constructed, rules of operation, and qualification tests for operating personnel to ensure that they know their jobs. Nuclear reactors must operate under a very high standard of quality assurance, requiring staff members to audit, evaluate, survey, and verify that all procedures and maintenance are being performed as they should be.

An important part of a safety system is strict adherence to design requirements: the reactor must have a negative power-reactivity coefficient; the safety rods must be injectable under all circumstances; and no single regulating rod should be able to add substantial reactivity rapidly. Another important design requirement is that the structural materials used in the reactor must retain acceptable physical properties over their expected service life. Finally, construction is to be covered by stringent quality-assurance rules, and both design and construction must be in accordance with standards set by major engineering societies and accepted by regulatory bodies.

Since no human activity can be shown to be absolutely safe, all these measures cannot reduce the risks to zero. However, it is the aim of the regulations and safety systems to minimize risks to the point where a reasonable individual would conclude that they are trivial. What this de minimis risk value is, and whether it has been achieved by the nuclear industry, is a subject of bitter controversy, but it is generally accepted that independent regulatory agencies—the United States' Nuclear Regulatory Commission (NRC), the United Kingdom's Office for Nuclear Regulation (ONR), the International Atomic Energy Agency (IAEA), and similar agencies around the world—are the proper judges of such matters.

The Windscale Accident of 1957

Prior to the development of current preventive-design philosophies, the world's first large-scale nuclear reactor accident took place in October 1957 at Windscale, Cumberland (now part of Cumbria), northwestern England. The Windscale plant was powered by a pair of identical reactors, known as Piles 1 and 2, that were an air-cooled, graphite-moderated design. Initially constructed between 1946 and 1950 to produce plutonium for nuclear weapons, they also provided energy for electricity production. However, with Britain fully engaged in the nuclear arms race, the reactors' operations were subject to direct political influence, and Windscale engineers were driven to modify the fuel design in order to increase plutonium and tritium production. These modifications increased heat generation in the fuel and therefore temperature levels within the reactor cores.

After approximately seven years of successful operation, the increased production rate became too much for Pile 1 to handle. Possibly as a result of imperfections in one of the fuel elements, the core began to overheat, and fire broke out within the lithium-magnesium-clad uranium fuel slug. In an attempt to remove heat from the core, operators turned on all the reactor's air-cooling fans. Unfortunately, this only fanned the flames and spread the fire through the core region. In addition, a significant amount of radioactive contamination was released through the ventilation stacks when the fans were turned on. Operators ultimately put out the fire by turning the fans off and forcing water through the core.

The Windscale event caused much less damage than the Three Mile Island, Chernobyl, and Fukushima accidents of later years. Nevertheless, it provided an explicit demonstration that political agendas (in this case, participation in the nuclear arms race) must be separated from those of energy and safety.

The Reactor Safety Study of 1972–75

In 1972, as part of an effort to evaluate the risks from nuclear power plants, the U.S. Atomic Energy Commission (a predecessor of the NRC) authorized a major safety study. Conducted with major assistance from a number of laboratories, the AEC's study involved the application of probabilistic risk assessment (PRA) techniques for the first time on a system as complex as a large nuclear power reactor. Also for the first time, the study compared the risk of a nuclear power plant accident with other events such as natural disasters and human-caused events. This work resulted in the publication in 1975 of a report titled Reactor Safety Study, also known as WASH-1400. The most useful aspect of the study was its delineation of components and accident sequences (scenarios) that were determined to be the most significant contributors to severe accidents.

The Reactor Safety Study concluded that the risks of an accident that would injure a large number of people were extremely low for the light-water reactor (LWR) systems being analyzed. This conclusion, however, was subject to very large quantitative uncertainties and was challenged. One basic problem with PRA techniques is that it cannot easily be confirmed by experience when the level of risk has been reduced to low values. That is to say, if PRA predicts that a reactor is subject to, say, one failure in 10,000 years, there is no way to prove that statement with only a few, or even with 10,000, years of experience. Thus, the results of the Reactor Safety Study as to risk levels were not confirmable. Nevertheless, updated versions of the report still provide the framework and reference for nuclear-related probability risk assessment.

Three Mile Island and Chernobyl

The principles established by the Reactor Safety Guide were given an unexpected test in 1979 when Three Mile Island Unit 2 near Harrisburg, Pennsylvania, suffered a severe accident. Through the failure of an important valve to operate correctly, cooling water to the core was lost, parts of the core were melted and the rest of it destroyed, and a large quantity of fission products was released from the primary reactor system to the interior of the containment structure. The equipment failure was exacerbated by reactor operator error, as the emergency core cooling system was deactivated by operator action because of a misinterpretation of the type of accident that was occurring. Fortunately, the containment vessel of the reactor building fulfilled its function, and only a small amount of radioactivity was released, demonstrating the wisdom of incorporating this structure into a rigorous design. All the same, a severe accident had occurred.

Aerial view of Three Mile Island near Harrisburg, Pennsylvania.

Many investigations of the Three Mile Island accident followed. Recommendations differed among them, but a common thread was that the human element was a much more important factor and higher risk to the safe operations of a nuclear power plant than had been recognized. The human element pertained not only to the operating staff but also to the management of nuclear plants and even the NRC itself. As a result, following the accident many changes in operator training and in technical and inspectorate staffing were implemented, just as a number of hardware enhancements were introduced. It is generally believed that these changes have been effective in reducing the likelihood of the occurrence of accidents as severe as that at Three Mile Island. As a side issue to this, however, the operating costs of nuclear power plants have escalated sharply as more and more highly trained people have been added to the operating staffs.

The significance of the human element, particularly as it relates to plant management and rigorous high-level regulatory decision making, was borne out again by the Chernobyl disaster of 1986. One of the four reactors in a nuclear power station approximately 100 km (60 miles) north of Kiev, Ukraine (then part of the Soviet Union), exploded and caught fire as the result of an ill-conceived experiment (a test to quantify how long the steam turbines would run while coasting to a stop if the reactor was abruptly shut down). Before the event was brought under control, an estimated 25 percent of the radioactive contents of the reactor had been released in a high cloud plume. Approximately116,000 people had to be evacuated, and a large area surrounding the plant received fallout so great that it could not be farmed or pastured as a result of this accident. Significant levels of radiation were detected as far north as Scandinavia and as far west as Switzerland.

In September 2005 the Chernobyl Forum, comprising seven United Nations organizations and programs, the World Bank, and the governments of Belarus, Russia, and Ukraine, published a three-volume, 600-page report assessing the impact of the accident on public health. Approximately 50 emergency workers had died of acute radiation sickness shortly after the accident, and 9 children had died from thyroid cancer because of radiation exposure. From among the 200,000 emergency workers who were present at the site in the first year following the accident, the people who were evacuated, and the 270,000 residents of the most heavily contaminated areas, an additional 3,940 people were likely to die from cancer during a prolonged period after the accident.

Investigation of the Chernobyl accident placed the largest blame, as with the Three Mile Island mishap, on poor management both at the plant and within the government bureaucracy. Because these accidents primarily resulted from human failings rather than from some intrinsic factor, most experts have continued to believe that nuclear energy can be a safe source of power. There is, however, a condition on the conclusion that nuclear power is by and large a safe form of power. The facilities for generating this power must be designed, built, and operated to high standards by knowledgeable, well-trained professionals, and a regulatory mechanism capable of enforcing these standards must be in place.

Mitigating Measures

Systems and Structures

Mitigating measures, also referred to as safety systems, are systems and structures that prevent accidents from proceeding to a catastrophic outcome in the event they do occur. Two of the principal mitigating measures, described in this topic Reactor design and components, are (1) the safety rod systems that quickly put the reactor into a subcritical state and prevent a supercritical accident and (2) the containment structure that prevents radioactive materials from being released into the atmosphere. Other significant mitigating measures include the emergency core-cooling system, whose purpose is to provide sufficient cooling of the core and fuel region within the vessel upon a loss of reactor coolant, and the emergency power system, which is designed to supply electrical power to support systems in the event that the normal supply is disrupted. Emergency power systems are necessary so that detectors, circulating pumps, valves, and other critical components continue to operate as necessary to remove decay heat. An extreme mitigating measure is the evacuation of personnel who might otherwise be heavily exposed in a reactor installation.

According to PRA studies, three categories of events are primarily responsible for the risks associated with LWRs—namely, station blackout, so-called transient without scram, and loss of cooling. In station blackout, a failure in the power line to which the station is connected is postulated. The proposed emergency defense is a secondary electrical system, typically a combination of diesel generators big enough to drive the pumps and a battery supply sufficient to run the instruments. In transient without scram, the assumed event is an insertion of positive reactivity—for example, through an undesired withdrawal of the shim rods. The protective safety system response in this case is the rapid and automatic insertion of the safety rods. In loss of coolant, the event is assumed to be caused by a mechanical failure of the normal cooling system such that a certain amount of the coolant is lost. The emergency response is activation of an emergency core-cooling system. In all such measures, proper operator action and proper functioning of the appropriate backup system are paramount aspects of emergency response.

Other reactor designs pose different types of risk. For example, neither the pool-type liquid-metal reactor (LMR) nor the high-temperature gas-cooled reactor (HTGR) is at major risk with regard to loss of coolant flow and perhaps not with regard to station blackout. However, the LMR and perhaps the HTGR are at some risk from events that might cause air or water to enter the coolant system. The hazard is that reactor materials, sodium or graphite, could chemically react with air and water, causing what is known as an exothermic reaction that releases large amounts of heat in addition to the decay heat already existing within the core region. The hazard is greater with sodium in the LMR than it is with graphite in the HTGR.

Fukushima

A failure of the main power line and a loss of backup power were at the heart of the second worst nuclear accident in the history of nuclear power generation (after Chernobyl)—a partial meltdown in 2011 at the Fukushima Daiichi ("Number One") plant in Japan. That facility, located on Japan's Pacific coast in northeastern Fukushima prefecture, was made up of six boiling-water reactors (BWRs) constructed between 1971 and 1979, three of which were operational and one of which was under maintenance, its fuel having been stored out of the core in the reactor's spent fuel storage pool. A powerful earthquake shook all units at the plant, initiating an automatic shutdown, or scram. Immediately after the earthquake, all safety systems in each unit were operable, though a few were slightly damaged. However, less than one hour after the earthquake, a tsunami struck the shoreline where the reactor units were built. The tsunami reached heights much greater than the reactors were designed to withstand, and ultimately it cut off the main power supply to the facility and damaged the backup generators by flooding their housing structures. Although the reactors withstood both an earthquake and a tsunami beyond their design requirement, the prolonged power outage drained backup batteries incorporated into the emergency core-cooling system, which led to a loss of capability to remove decay heat. Despite the best efforts of the reactor operators and emergency responders, rising temperatures within each reactor's core eventually caused a partial meltdown of the fuel rods, a fire in the storage reactor, explosions in the outer containment buildings (caused by a buildup of hydrogen gas), the release of radioactive steam into the air, and the leakage of radioactive water into the ocean. As workers struggled to cool and stabilize the three cores by pumping seawater and boric acid into them, government officials established a 30-km (18-mile) evacuation zone around the plant. Approximately one month after the initiating event, the reactor cores were stabilized, cracks in the foundations of the containment vessels were sealed, and irradiated cooling water began to be pumped to a storage building until it could be properly treated.

Two of the damaged containment buildings at the Fukushima Daiichi nuclear power plant, northeastern Fukushima prefecture, Japan, several days after the March 11, 2011, earthquake and tsunami that crippled the installation.

The Fukushima accident made it all too clear that another type of risk can arise from external events: earthquakes and tsunamis may not be two separate events but rather be two successive events in which an earthquake will cause structural damage to a reactor and will also initiate a tsunami. The risk associated with an earthquake of plausible magnitude is minimized by building plants away from faults and by making use of earthquake-resistant mechanical design and construction features. Furthermore, the addition of dikes and water barriers reduces the risk of damage by a tsunami. Added construction features such as water barriers must be able to withstand both an earthquake and a tsunami, as these are likely to be coupled events.

In contrast to the Three Mile Island and Chernobyl accidents, which were largely blamed on staffing issues, the "weak link" in the Fukushima accident seemed upon immediate observation to be the physical plant itself rather than human error. However, because the plants were not designed to handle the natural disaster that took place, fault can be found with the design process, in a sense pointing out human error once again as the most failure-prone component in the nuclear industry.

Emergency Response

Each regulating body that oversees the operations of a country's nuclear power has its own methods for identifying and responding to emergency conditions. In the United States, the NRC has an emergency classification system that identifies four levels of severity in conditions at a nuclear power plant:

A man is checked for radiation exposure after having been evacuated from the quarantine area around a nuclear power station in Fukushima prefecture, Japan, that was damaged in the March 11, 2011, earthquake and tsunami.

1. Notification of unusual events: Potential degradation in the level of safety of the plant, but no release of radioactive material requiring off-site response or monitoring.
2. Alert: Actual or potential substantial degradation in the level of safety of the plant, with a release of radioactive material from the plant expected.
3. Site area emergency: Actual or likely major failures of plant functions needed for protection of the public, with radioactivity levels potentially above acceptable thresholds at the boundary of the power plant.
4. General emergency: Actual or imminent substantial core damage or melting of reactor fuel with the potential for loss of containment integrity; radioactive material is released and may be above acceptable thresholds beyond the boundary of the power plant.

On a worldwide scale, the IAEA has developed the International Nuclear and Radiological Event Scale (INES), to be applied to any event occurring in the agency's signatory states that is associated with nuclear facilities and with the transport or storage of nuclear materials and radiation sources. The INES offers a common event scale for all parties that interact with nuclear power or radiological sources in any part of the world. The scale includes seven independent event levels; the lower three are referred to as "incidents" and the upper four as "accidents." A declaration of a specific level is determined by identifying specific criteria that have an impact on defense-in-depth of the nuclear power plant, radiological barriers and controls, and people and the environment. The seven levels and some of the important criteria are as follows:

1. Anomaly: Minor problems with safety components, with significant defense-in-depth remaining.
2. Incident: Significant contamination within the facility into an area not expected by design, with exposure of a worker in excess of the statutory annual limits.
3. Serious incident: Severe contamination in an area not expected by design, with a nonlethal health effect such as a burn on a worker from radiation.

4. Accident with local consequences: Fuel melt or damage to fuel resulting in more than 0.1 percent release of core inventory; release of significant quantities of radioactive material within an installation, with a high probability of significant public exposure and at least one death from radiation.
5. Accident with wider consequences: Severe damage to reactor core; release of large quantities of radioactive material within an installation, with a high probability of significant public exposure and several deaths from radiation.
6. Serious accident: Significant release of radioactive material likely to require implementation of planned countermeasures.
7. Major accident: Major release of radioactive material with widespread health and environmental effects requiring implementation of planned and extended countermeasures.

Under the INES, Three Mile Island is classified as Level 5, an accident with wider consequences, whereas both Fukushima and Chernobyl are Level 7, major accidents.

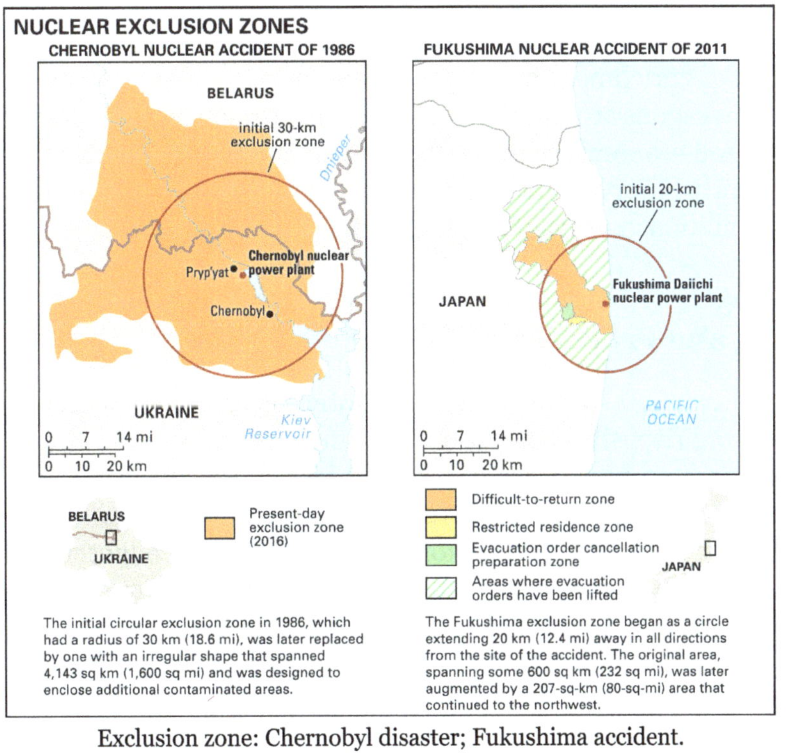

Exclusion zone: Chernobyl disaster; Fukushima accident.

The Nuclear Fuel Cycle

No discussion of nuclear power is complete without a brief exposition of the nuclear fuel cycle. The whole point of a reactor is, after all, to initiate and control the process of fission on a very large scale in nuclear fuel, and the low cost of fueling is the chief reason for the economic competitiveness of nuclear power. The principal steps of the fuel cycle include uranium mining and extraction from its ore (processing), uranium enrichment, fuel fabrication, loading and irradiation in the reactor (fuel management), unloading and cooling, reprocessing, waste packaging, and waste disposal.

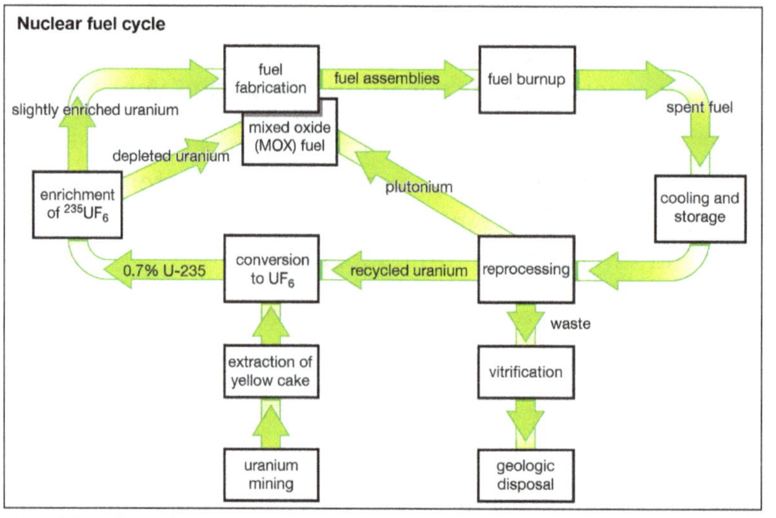

A "closed loop" nuclear fuel cycle, showing the reprocessing of uranium-235 and plutonium from spent fuel for use in new fuel assemblies.

The nuclear fuel cycle also is an integral step in the production of plutonium for nuclear weapons, and the technologies of enrichment and reprocessing in particular have been key factors in the proliferation of these weapons around the world. For this reason and also for a host of other political, environmental, and economic reasons, the various steps in the nuclear fuel cycle are closely regulated and frequently observed under terms of international treaties. Conflicts between some countries' nuclear ambitions and various international conventions have sometimes generated great controversy.

Uranium Mining and Processing

Uranium is extracted from ores whose uranium content is often less than 0.1 percent (one part per thousand). Most ore deposits occur at or near the surface; whether they are mined through open-pit or underground techniques depends on the depth of the deposit and its slope. The mined ore is crushed and the uranium chemically extracted from it at the mouth of the mine. The residue remains naturally radioactive, as it contains long-lived radioactive daughter nuclei of uranium and has to be carefully managed to minimize the release of radioactive contaminants into the environment. The uranium concentrate, which is known as yellow cake, consists of uranium compounds (typically 75 to 95 percent). It is shipped to a chemical plant for further purification and chemical conversion.

Enrichment

Several enrichment techniques have been developed, though only two of these methods are used on a large scale; these are gaseous diffusion and gas centrifuging. In gaseous diffusion, natural uranium in the form of uranium hexafluoride gas (UF_6), a product of chemical conversion, is encouraged (through a mechanical process) to seep through a porous barrier. The molecules of $^{235}UF_6$ penetrate the barrier slightly faster than those of $^{238}UF_6$. Since the percentage of ^{235}U increases by only a very small amount after traversal of the barrier, the process must be repeated over and over in thousands of stages to obtain the necessary enrichment for commercial nuclear power use.

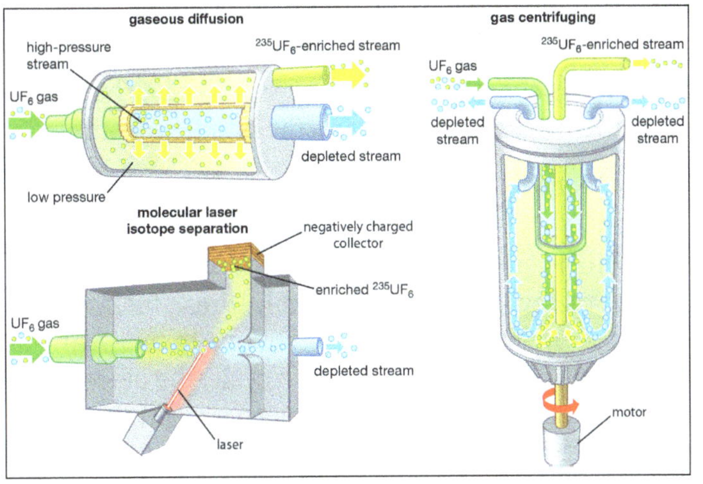

Three uranium-enrichment processes.

In gas centrifuging, the UF_6 gas is fed into a high-speed centrifuge. The centrifuge is balanced very well at the top bottom and spins at an extremely high rate. Because of the relative centripetal forces that each atom experiences, the lighter species of this mixture of gaseous molecules, including ^{235}U, tend to concentrate near the centre of the spinning centrifuge, while the heavier ones accumulate along the wall. These mixtures are then siphoned off. The degree of enrichment per stage in a centrifuge is greater than that obtained in a gaseous diffusion chamber, and the process uses less energy than gaseous diffusion does, but centrifuges are more expensive pieces of equipment.

An experimental enrichment method with much commercial potential is laser separation. This process is based on the principle that isotopes of different molecular weight absorb light of different frequencies. Once a specific isotope has absorbed radiation and has reached an excited state, its properties may become quite different from the other isotopes; it is then separated on the basis of this difference. In one method known generically as MLIS (molecular laser isotope separation)—or commercially as SILEX (separation of isotopes by laser excitation)—gaseous UF_6 is exposed to high-powered lasers tuned to the correct frequencies to cause the molecules containing ^{235}U (but not ^{238}U) to lose electrons. In this (ionized) form, the ^{235}U-containing molecules are separated from the stream on the basis of their different electric charge. Proponents of laser separation claim that the method consumes less energy and wastes less starting material than, for example, gaseous diffusion.

Fabrication

This step involves the conversion of the suitably enriched product material to the chemical form desired for reactor fuel. The only fuel fabricated on a large scale is for light-water reactors (LWRs).

The chemical form prepared for the LWR is uranium dioxide. Produced in the form of a ceramic powder, this compound is ground to a very fine flourlike consistency and inserted into a die, where it is pressed into a pellet shape—in the case of some LWR fuels, approximately 6 mm in diameter and 10 mm in length (that is, about 0.25 × 0.4 inch). Next the pellet is sintered in a furnace at 1,500–1,800 °C (approximately 2,700–3,300 °F). This sintering, similar to the firing of other ceramic ware, produces a dense ceramic pellet. The pellets are loaded into prefabricated zirconium alloy cladding tubes, which are then filled with an inert gas and welded shut. Once the zirconium alloy tubes have been sealed, they go through significant testing to verify that there are no leaks. These tubes, called rods or pins, are then bundled together with proper spacing ensured by top and bottom manifolds through which the ends of the pins pass as well as spacer grids distributed along the middle portion of the pins. Together with other necessary hardware, the bundle constitutes a fuel assembly.

Fuel Management

Fuel is loaded into a reactor in a very specific and well-controlled pattern so as to obtain the most energy production before the material becomes unusable. Fresh fuel is more reactive than old fuel. Typically, a reactor is fueled in cycles, each cycle lasting one to two years, and a fuel batch is kept in the reactor for three or four cycles. At the end of each cycle, the oldest fuel is removed—normally this consists of about one-third the fuel content in the core—and fresh fuel loaded. The partially burned fuel that remains, however, is shuffled before the fresh fuel is installed. The objective of this procedure is to achieve a fuel assembly arrangement of maximum reactivity while keeping the power distribution among the different fuel assemblies as even as possible and within technical specifications.

Fuel burnup—that is, energy production—is limited by two factors. After significant burnup has occurred, the physical properties of the fuel become degraded, and it is not prudent to continue to keep it in the reactor. Also, after some burnup, the old fuel no longer contributes useful reactivity to the reactor. The fuel design, including its initial enrichment, is such that these two limits are made to coincide approximately.

Unloading and Cooling

Spent reactor fuel is extremely radioactive, and its radioactivity also makes it a source of heat. When the spent fuel is removed from the reactor, it must continue to be both shielded and cooled. This is accomplished by placing the spent fuel in a water-storage pool, or spent-fuel cooling pool, located next to the reactor. The water in the pool contains a large amount of dissolved boric acid, which is a strong absorber of neutrons; this ensures that the fuel assemblies in the pool do not go critical. (Pool water is also a common source of emergency cooling water for the reactor.)

Spent nuclear fuel rods stored underwater at the Bruce Power site near Tiverton, Ontario, Canada.

Pools vary in size; older spent-fuel cooling pools are able to accommodate only about 10 years' worth of spent fuel. As the pools fill up, more spent fuel storage is needed. Additional storage space can be gained by loading spent fuel into the pool more densely than originally planned, by building a new pool, or by removing the oldest fuel assemblies from the existing pool and storing them in air-cooled concrete and steel silos—called spent-fuel storage casks—located aboveground. This last method becomes feasible after fuel has been stored in cooling pools for two or three years, because radioactivity and the rate of heat generation decrease rapidly over this period. Dense storage in existing pools and casks tends to be less expensive and more economical for utilities than building new pools.

Reprocessing

Both the converted plutonium and residual uranium-235 in spent fuel can be recycled by chemically reprocessing the fuel and extracting the specific elements of interest. Reprocessing not only provides a means to recycle nuclear fuel, but it also can reduce the volume and radioactivity of the waste material that must ultimately be eliminated by some method of permanent disposal.

One motivation for reprocessing is ultimately to provide a "closed-loop" fuel cycle within the nuclear industry. Closed-loop refers to recycling with 100-percent efficiency of all materials that are fabricated for use as nuclear fuel (including the most commonly used fuel, uranium dioxide pellets). Though the goal of 100-percent efficiency has yet to be attained by any country's nuclear industry, a closed-loop fuel cycle is not an unrealistic ambition, based on current progress in reprocessing technology. Many benefits would result from fuel recycling, including lower cost for fuel (once the recycling infrastructure was in place) and reduced quantities of spent fuel to be stored on reactor sites around the world.

Reprocessing Methods

The most common method for reprocessing, known as the PUREX (for plutonium-uranium extraction) process, begins with dissolving the spent fuel in nitric acid and contacting the acid solution with oil in which tributyl phosphate (TBP) has been dissolved. TBP is a complexing agent for uranium and plutonium, forming compounds with them that bring them into the oil solution. A physical separation of the (immiscible) oil and acid serves to remove the desired products from the nitric acid solution (which still contains all the fission products). The uranium and plutonium are then washed out of the TBP back into a water solution and separated from each other by various means to the degree desired. Thus, reprocessing produces three product streams: (1) a purified uranium product, (2) a plutonium product that may be either pure or mixed with uranium, and (3) a waste stream of fission products dissolved in nitric acid.

Reprocessing Policies

During the period of ambitious nuclear power plant construction in the United States in the 1950s and '60s, it was generally assumed that after two to five years, spent fuel would be delivered to a reprocessing plant. Some commercial reprocessing plants were built or planned, but by the mid-1970s the cost of reprocessing had escalated to a point where its economics became questionable. Also, in 1977 Pres. Jimmy Carter, in order to take a public, symbolic stand against nuclear proliferation, declared that the federal government would permanently defer all permits for the commercial reprocessing and recycling of plutonium. Carter's directive was rescinded by his successor, Ronald Reagan, and it has not been reinstated by any subsequent president. Even so, reprocessing is still not done commercially in the United States, partly because of the huge costs of building a reprocessing plant in a period when the supply of uranium ore has been sufficient to satisfy demand relatively cheaply.

Policy and institutional arrangements have been different in France and the United Kingdom, where commercial plants reprocess spent fuel not only from nuclear plants in the host countries but also from plants in other countries. The reprocessed plutonium can be used not only as fuel for proposed future liquid-metal reactors (LMRs) but also to help fuel existing LWRs. In the latter application, the plutonium is utilized in mixed oxide (MOX) form—a combination of uranium and plutonium dioxides having 3 to 6 percent plutonium.

A number of other countries, including Russia, India, Japan, and China, reprocess their spent fuel or plan to do so. The proactive reprocessing efforts of these countries have reduced the waste scheduled for long-term disposal to amounts well below those that are accumulating in countries that do not reprocess.

Waste Disposal

In the absence of reprocessing, spent fuel is considered to be waste and must be prepared for permanent disposal in a separate facility. In addition, the waste stream from spent-fuel reprocessing must also be disposed of. Many nuclear countries, from the United States to China to Finland, have researched the technologies and geologic locations for disposal sites, but no permanent disposal site is in use anywhere in the world. Pending approval and construction of disposal sites, all spent fuel and processed waste are being kept either in cooling pools or in aboveground storage casks.

Containers of nuclear waste stored above ground at the Diablo Canyon Power Plant, San Luis Obispo county, California.

Waste Conditioning

Spent fuel must be sealed in containers that are expected to remain viable in stable (and presumably underground) disposal sites over centuries and even millennia. Though no permanent disposal site is currently operational, the preparing, or conditioning, of spent fuel for disposal is expected to follow the same basic process. After storage aboveground for one to five years, the fuel pins are to be removed from their assemblies. End manifolds and components within the fuel assembly that contain no fuel are to be removed and the pins repacked into a dense lattice emplaced in a corrosion-resistant stainless-steel canister. A cover is to be welded on and the canister covered with an overpack.

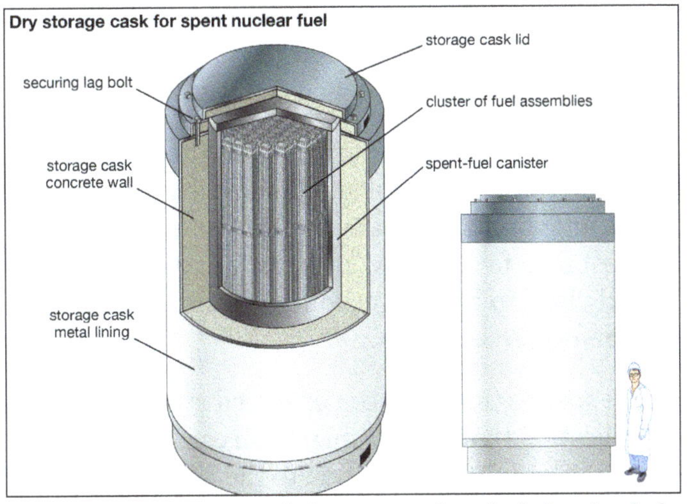

Cutaway diagram of a dry storage cask for spent nuclear fuel, showing fuel assemblies packed into a metal canister that is encased in a concrete cask.

Some waste is generated every year in the form of the fission-product solution that arises from reprocessing. (Reprocessed fuel from nuclear-weapons production reactors also generates this type of waste.) One glassmaking process for conditioning this waste is operational on an industrial scale in France, the United Kingdom, and Japan and has been tested in many other countries. The waste solution is completely evaporated, leaving behind the fission products in the solid residue, which is heated until all the constituent nitrate salts have been converted to oxides. These oxides are then put into a glass-forming oven and mixed with materials that will produce a borosilicate glass (similar to the commercial glass known as Pyrex). The fission-product oxides dissolve in the glass as it forms. The glass melt is subsequently poured into a steel canister, 200–400 mm (8–16 inches) in diameter and approximately 1 metre (40 inches) high, where it solidifies. Once covered with an overpack of bentonite clay (for shielding, molecular diffusion, and chemical isolation), the solid canister-like block is ready for disposal.

Geologic Disposal

The waste-disposal method currently being planned by all countries with nuclear power plants is called geologic disposal. This means that all conditioned nuclear wastes are to be deposited in mined cavities deep underground. Shafts are to be sunk into a solid rock stratum, with tunnel corridors extending horizontally from the central shaft region and tunnel “rooms” laterally from the corridors. The waste would be emplaced (by remotely controlled or robotic devices) in holes drilled into the floors of these rooms, after which the boreholes would be sealed and the rooms and corridors backfilled. When the entire operation has been completed (perhaps after approximately 30 years of operation), the shafts too would be backfilled and sealed.

Risks of Nuclear Waste Disposal

When a holistic view is taken of the nuclear waste-disposal process, the risks seem extremely small, yet among the general public, these risks are one of the most-feared aspects of the nuclear fuel cycle. Nuclear waste retains its very intense level of radioactivity for several hundred years, but after a thousand years have passed, the remaining radioactivity, while persistent, is at a level comparable to (though still greater than) that of an equivalent quantity of natural uranium ore.

This separates the safety problem into two time periods: a first millennium during which it is crucial to ensure tight retention of the wastes in the repository, and a subsequent period during which it is necessary only to ensure that any release that occurs is small and slow.

One possible route for the emergence of radioactive waste to the surface would be the impingement of groundwater into the underground disposal site, followed by corrosion of the waste canisters, dissolving of waste material, and discharge of the resulting solution to the environment. However, water migrates slowly in most rock formations, and, contrary to the popular belief that any dissolution of waste would quickly lead to high-level contamination, only low levels of contamination are projected in even the worst-case scenarios.

The migration of radioactive species that has been observed at shallow burial sites for low-level radioactive waste is not an indication that similar migration can be expected for high-level waste located in a repository deep underground. In addition to the near insolubility of the waste material, waste form engineering, particularly of corrosion-resistant containers, provides extra protection against such dispersal. Moreover, most of the dispersal problem in shallow disposal sites is caused by biochemical products that would not exist in deep formations.

Finally, a great deal of care is to be expended in selecting the site of the repository. Various conditions are mandatory: the repository must not be near a populated area; the rock stratum selected must be deep (300 metres [1,000 feet] or more) and, as much as possible, naturally sealed from aquifers; and any discharge of the water table into the surface waters should be slow. Furthermore, the site must be in a tectonically inactive zone so that earthquakes will not break the seal.

The risk of high-level waste burial is almost certainly smaller than the risks of reactor accidents and even smaller than the risks arising from improperly managed mine tailings. Nonetheless, the siting of a repository must be handled with political sensitivity, and the confirmation of acceptable hydrologic and geologic conditions must have a high degree of validity. There are many acceptable sites in principle, but confirming acceptability for any one of them is a large and expensive technical undertaking. Both politically and technically, site selection is probably the biggest problem in the nuclear fuel cycle, causing the postponement, revision, and even termination of many siting efforts around the world.

Reactor Development

Since the inception of nuclear power on an industrial scale in the mid-20th century, fundamental reactor designs have progressed so as to maximize efficiency and safety. In this historical progression, four distinct reactor generations can be discerned. Generation I reactors were the first to produce civilian nuclear power—for example, the reactors at Shippingport in the United States and Calder Hall in the United Kingdom. Generation I reactors have also been referred to as "early prototypic reactors." The mid-1960s gave birth to Generation II designs, or "commercial power reactors." Most nuclear power plants in operation today employ Generation II technology.

Generation II designs incorporated a number of elements to increase the safety of the reactor and decrease the risks associated with accidents. However, the Generation II elements are considered to be "active safety systems"; that is, they must be activated by human controllers and cannot operate if electric power systems are shut down. In an effort to advance safety even further, a new generation of "advanced light-water reactors" was designed beginning in the mid-1990s. These Generation

III designs incorporate so-called passive safety systems into the reactor structure. Passive systems are intended to increase reactor safety by operating with no human intervention or electrical power. Two prominent Generation III designs are the European Pressurized Water Reactor (EPR) and the Westinghouse Advanced Plant 1000 (AP1000) pressurized water reactor. In the AP1000 design, in the event of a complete loss of electrical power (including emergency backup generators), control rods would drop into the reactor core, immediately stopping the nuclear chain reaction, and continuing decay heat would be transferred out of the reactor containment by a system of gravity-fed cooling tanks. One tank, located inside the sealed containment structure, would feed water to the core; this water would boil and rise as steam to the top of the containment structure, where it would condense and flow back to the internal cooling system. The heat of condensation in turn would be transferred to the containment structure, which would be cooled by water flowing by gravity from an external tank located atop the containment. Water evaporating on the exterior of the containment would complete the transfer of reactor heat to the atmosphere, where it would dissipate.

The nuclear industries of several countries are currently planning Generation IV nuclear power plants, or "next generation nuclear plants" (NGNPs), which are designed with the intent to be built starting in the second quarter of the 21st century. For a reactor to be categorized as an NGNP, it would have to satisfy several requirements, including (1) being highly economical, (2) incorporating enhanced safety, (3) producing minimal waste, and (4) being proliferation resistant. One NGNP concept is the very high temperature reactor (VHTR), a helium-cooled, graphite-moderated reactor using a variety of fuels that would create enough heat to generate electricity and also supply other industrial processes, such as the production of hydrogen from water.

The First Atomic Piles

Soon after the discovery of nuclear fission was announced in 1939, newspaper articles reporting the discovery mentioned the possibility that a fission chain reaction could be exploited as a source of power. World War II, however, began in Europe in September of that year, and physicists in fission research turned their thoughts to using the chain reaction in an atomic bomb. In the United States, Pres. Franklin D. Roosevelt was persuaded by a letter from Albert Einstein to initiate a secret project devoted to this purpose. The Manhattan Project included work on uranium enrichment to procure uranium-235 in high concentrations and also research on reactor development. The goal was twofold: to learn more about the chain reaction for bomb design and to develop a method of producing a new element, plutonium, which was expected to be fissile and could be isolated from uranium chemically.

Reactor development was placed under the supervision of the leading experimental nuclear physicist of the era, Enrico Fermi. Fermi's project began at Columbia University and was first demonstrated at the University of Chicago, centred on the design of a graphite-moderated reactor. On December 2, 1942, Fermi reported having produced the first self-sustaining chain reaction. His reactor, later called Chicago Pile No. 1 (CP-1), was made of pure graphite in which uranium metal slugs were loaded toward the centre with uranium oxide lumps around the edges. This device had no cooling system, as it was expected to be operated for purely experimental purposes at very low power (roughly 10 kilowatts of thermal energy). CP-1 was subsequently dismantled and reconstructed at a new laboratory site in the suburbs of Chicago, the original headquarters of what is now Argonne National Laboratory. The device saw continued service as a research reactor until it was finally decommissioned in 1953.

Power output is thermal except where noted as megawatts (e), signifying electrical.

On the heels of the successful CP-1 experiment, plans were quickly drafted for the construction of the first production reactors (for producing the plutonium to be used in the atomic bomb). These were the early Hanford, Washington, reactors, which were graphite-moderated, natural uranium-fueled, water-cooled devices. As a backup project, a production reactor of air-cooled design was built at Oak Ridge, Tennessee. When the Hanford facilities proved successful, this reactor was completed to serve as the X-10 reactor at what is now Oak Ridge National Laboratory. The first enriched-fuel research reactor was completed at Los Alamos, New Mexico, in 1944 as enriched uranium-235 became available for research purposes. All of these efforts culminated in Trinity, the first test of an atomic explosive device, which took place on July 16, 1945, at Alamogordo, New Mexico.

Even before the war, it had been recognized that heavy water was an excellent neutron moderator and could be easily employed in a reactor design. During the Manhattan Project, this possible design feature was assigned to a Canadian research team, since heavy-water production facilities already existed in Canada. In late 1945, shortly after the end of the war, the Canadian project succeeded in building a heavy-water-moderated, natural uranium-fueled research reactor, the so-called ZEEP (Zero-Energy Experimental Pile), at Chalk River, Ontario.

Zero-Energy Experimental Pile.

Because of a lack of information on uranium-235 separation techniques, the first British efforts, which took place after the war, were centred on the use of natural uranium as a fuel. In 1947 GLEEP (Graphite Low Energy Experimental Pile), an air-cooled reactor with a graphite moderator and uranium metal fuel clad in aluminum, was constructed and went critical at Harwell, Berkshire, England, generating 100 kilowatts of thermal energy. The following year, a French reactor of similar power, known as EL-1 (for "heavy water 1") or Zoé (for "zero power, uranium oxide, heavy water"), was built at Châtillon, near Paris. The French reactor too used nonenriched uranium in its fuel.

In 1943 the Soviet Union began a formal research program to create a controlled fission reaction, explore isotope separation, and investigate atomic bomb designs. After the war, the program began to make significant progress toward the design of a fission weapon; in tandem, reactors were designed for the purpose of producing weapons-grade plutonium. The first Soviet chain reaction took place in Moscow in late 1946, using an experimental graphite-moderated natural uranium pile known as F-1. The first plutonium production reactor became operational at the Chelyabinsk-40 complex in the Ural Mountains of Russia in 1948.

From Production Reactors to Commercial Power Reactors

The earliest U.S. nuclear power project had been started in 1946 at Oak Ridge, but the program was abandoned in 1948, with most of its personnel being transferred to the naval reactor program. In 1953 the first prototype submarine reactor was started up (leading to the launching the next year of the first nuclear-powered submarine, the Nautilus), and also in 1953 Pres. Dwight D. Eisenhower announced the Atoms for Peace program. Atoms for Peace established the groundwork for a formal U.S. nuclear power program and also expedited international cooperation on nuclear power. The U.S. nuclear power program was devoted to the development of several reactor types. Three of these types ultimately proved successful in the sense that they remain as commercial reactor types today or as systems scheduled for future commercial use. These are the fast breeder reactor (now called the liquid-metal reactor, or LMR), the pressurized-water reactor (PWR), and the boiling-water reactor (BWR).

The first LMR was the Experimental Breeder Reactor, EBR-I, which was designed at Argonne National Laboratory and constructed at what is now the Idaho National Laboratory near Idaho Falls, Idaho. EBR-I was an early experiment to demonstrate breeding, and in 1951 it produced the first electricity from nuclear heat. A much larger experimental breeder, EBR-II, was developed and put into service (with power generation) in 1963.

The principle of the BWR was first demonstrated in a research reactor in Oak Ridge, but development of this reactor type was also assigned to Argonne, which built a series of experimental systems designated BORAX in Idaho. In 1955 one of these, BORAX-III, became the first U.S. reactor to put power into a utility line on a continuous basis. A true prototype, the Experimental Boiling Water Reactor, was commissioned in 1957. The principle of the PWR, meanwhile, had already been demonstrated in naval reactors, and the Bettis Atomic Power Laboratory of the naval reactor program was assigned to build a civilian prototype at Shippingport, Pennsylvania. This reactor, the largest of the power-reactor prototypes, went online in 1957; it is often hailed as the first commercial-scale reactor in the United States.

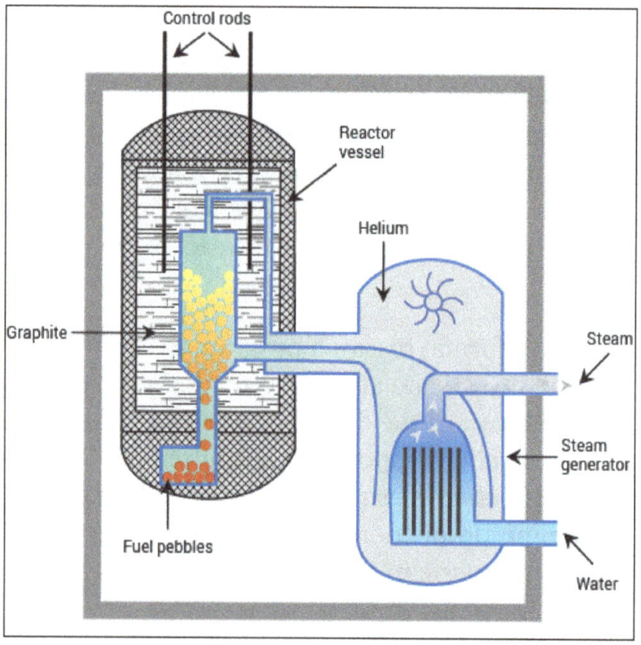

In 1949 the Soviet Union's nuclear efforts reached fruition with its first atomic bomb test. A decade later, the Soviets put their first nuclear-powered surface vessel, the icebreaker Lenin, into service, in conjunction with the United States' launch of its first nuclear-powered surface warship, the cruiser Long Beach. Having established their presence among the nuclear powers, the Soviets directed considerable efforts toward the use of nuclear energy to generate electric power through a standard steam cycle. In 1954 a graphite-moderated plutonium production reactor was modified in Obninsk, Russia, to create the first nuclear-powered electricity generator in the world. In the mid-1960s the Soviet Union commissioned two graphite-moderated 100-megawatt BWRs. Immediately following, reactors began to be constructed throughout the Soviet Union, the nominal power ratings increasing steadily. In 1973 the first 1,000-megawatt Reaktor Bolshoy Moshchnosti Kanalny (RBMK; "high-power channel reactor") went critical. The RBMK was an unusual pressurized-water design, originally intended to produce plutonium as well as generate electricity, in which the core was moderated by graphite. The Soviet nuclear power industry also built a more conventional PWR design, the Vodo-Vodyanoy Energetichesky Reaktor (VVER; "water-cooled water-moderated power reactor"), which, like U.S. PWR designs, developed from early work done on naval power plants.

In the United Kingdom, GLEEP was followed in the late 1940s and early 1950s by construction of the air-cooled Windscale No. 1 and No. 2 reactors, which produced plutonium for Britain's nascent nuclear weapons program. (Britain's first atomic bomb test took place in 1952.) Across the Calder River from Windscale, a new reactor, Calder Hall A, made history in 1956 by producing the world's first electric power generation on a commercial scale (while also producing plutonium for weapons). The Calder Hall reactors were cooled by compressed carbon dioxide gas and used a fuel of natural uranium metal sheathed in a new magnesium-alloy cladding. Because of the cladding, Calder Hall-type reactors were also known as Magnox reactors. Continuing from this technology, Britain went on to develop the Advanced Gas-Cooled Reactor (AGR), which used a fuel of enriched uranium dioxide, thus allowing higher reaction temperatures and more efficient power generation. The prototype AGR was built at Windscale and went critical in 1963. Commercial AGRs were built in the United Kingdom through the 1980s.

The Calder Hall nuclear power station.

France followed a modified British model with the G1, G2, and G3 graphite-moderated, gas-cooled reactors, which first went critical at Marcoule, on the Rhône River in southern France, in 1956–59. These reactors powered electric generators while also producing plutonium for France's nuclear weapons program. However, after initially focusing on gas reactors, France shifted to the development of light-water reactors. Ultimately, French commercial nuclear power plants were standardized on three basic PWR designs.

Canada embraced its abundant uranium ore reserves as well as its established ability to mass produce heavy water through the Canada Deuterium Uranium (CANDU) reactor design, which operates on natural or very slightly enriched uranium moderated by heavy water (deuterium oxide). The first CANDU reactor—producing about 20 megawatts of electricity—went critical in 1962 at Rolphton, Ontario, on the Ottawa River. Canada's first commercial nuclear power plant, located at Pickering, Ontario, on the shore of Lake Ontario, went into operation in 1971. By the end of the century, more than 20 CANDU reactors had been built in Canada, and many more had been built abroad.

Up to the 1970's, all commercial-scale nuclear power plants utilized thermal neutrons as their primary mechanism for fission. However, in 1973 the Soviet Union demonstrated the first commercial fast neutron reactor, built on the shore of the Caspian Sea near what is now Aqtau, Kazakhstan. The reactor, in addition to providing heat energy for a 120-megawatt electric power plant, also utilized its heat for desalination of water from the Caspian—a novel concept for alternative uses of nuclear energy that was subsequently adopted in other parts of the world. A number of fast neutron reactor prototypes had previously been developed in the United States, the United Kingdom, and elsewhere in the Soviet Union; however, no country had yet scaled this technology up to an economically useful level.

Growth of Nuclear Programs

Of the prototype commercial nuclear power plants that were built in the United States during the late 1950s and early 1960s, the most successful types used the light-water reactor system, either PWR or BWR. From the mid-1960s, larger units were ordered in the expectation of an ever-increasing commercial utilization of nuclear power, and by the early 1970s, nuclear plant orders were coming in at such a rapid pace that the unit sizes were increased so as to reduce the number of separate projects for which each vendor would have to provide trained staff. By the later years of the decade, however, the surfeit of orders in the United States had been followed by a large number of project cancellations, partly as a result of sharp reductions in estimates of how quickly the demand for electricity was expected to grow in the future. It began to appear that the new plants simply were not needed. Moreover, the cost of new nuclear plants had begun to escalate to the point where their economics became questionable. Finally, public fears of nuclear power, which had always been a factor, were brought to a head by the Three Mile Island accident of 1979. Following that event, not a single new reactor was approved in the United States until 2012, when the Nuclear Regulatory Commission approved the construction of two PWRs in Augusta, Georgia. By that time, a growing demand for electricity, as well as improvements in reactor designs and government incentives to install power plants that were not dependent on fossil fuels, had created the potential for a rebirth of nuclear power in the United States.

Numerous other countries around the world adopted nuclear power as a part of their energy strategy in the late 20th century. Most programs utilized light-water reactor technology of the BWR or PWR type. France, following the oil shocks of the 1970s, embraced nuclear power as a strategic energy source, to be pursued for reasons of national security as well as energy independence, and as such the country invested heavily in the technology rather than relying on U.S. designs. The PWR fleet developed by the French nuclear industry eventually succeeded in producing more than three-quarters of the country's electric power. Toward the end of the century, several European reactor vendors, including France's Areva and Germany's Siemens, produced an advanced Generation III PWR design known as the European Pressurized Reactor or Evolutionary Power Reactor (EPR). The standard EPR design yielded approximately 1,650 megawatts of electric power; early in the 21st century, construction began on several units in Europe as well as in China.

The Civaux nuclear power plant, using pressurized-water reactors, near Poitiers, western France.

Japan, a country that was particularly dependent on foreign imports for traditional fossil fuels, became an active member of the nuclear research community in the early 1960s; it began to harness nuclear power as a source of electricity soon after. Its first reactor to produce electricity was started in 1963, and its first commercial reactor, a gas-cooled British design, started operation in 1966. Following this initial construction, Japan switched to light-water reactors. By 2011, the year of the Fukushima accident, Japan had constructed more than 50 BWRs and PWRs for commercial power production and was generating almost one-third of its electricity from nuclear power plants. The first light water reactors in Japan were American designs, but Japan went on to develop its own advanced BWRs and PWRs. Japan was also developing facilities to reprocess its own spent reactor fuel instead of transporting the fuel to plants in Europe. In the aftermath of the Fukushima accident, all reactors in Japan, as they shut down for maintenance, were not permitted to restart unless they were able to pass new safety and stress tests.

The Ikata nuclear power plant, employing pressurized-water reactors, located in Ehime prefecture, Shikoku, Japan.

As part of its nuclear weapons program, China developed reactor and reprocessing technology in the 1950s and 1960s, at first with Soviet assistance but soon using its own considerable resources. By the first decade of the 21st century, China had grown to become the world's largest consumer of energy, and the country's leaders had promulgated a strategic energy plan that included a doubling of nuclear power generation in the period 2005–20. Fourteen nuclear power plants were in operation and more than 25 under construction, and many more were being planned. The first commercial-scale nuclear power plant in China had started operation in 1994 at Daya Bay on the coast of Guangdong province, near Hong Kong. The Daya Bay reactors were PWRs of French design, and PWRs formed the basis of China's growth into the 21st century—though CANDU reactors were also built, and both high-temperature gas-cooled reactors (HTGRs) and fast-neutron liquid-metal reactors (LMRs) were under consideration for the longer-term future. As a part of its strategic energy effort, China, like Japan, developed reprocessing technology in order to create a closed-loop nuclear fuel cycle.

The Tianwan nuclear power plant, using pressurized-water reactors, in Lianyungang, Jiangsu province, China.

Breeder Reactor

A breeder reactor is a nuclear reactor that generates more fissile material than it consumes. Breeder reactors achieve this because their neutron economy is high enough to create more fissile fuel than they use, by irradiation of a fertile material, such as uranium-238 or thorium-232 that is loaded into the reactor along with fissile fuel. Breeders were at first found attractive because they made more complete use of uranium fuel than light water reactors, but interest declined after the 1960s as more uranium reserves were found, and new methods of uranium enrichment reduced fuel costs.

Fuel Efficiency and Types of Nuclear Waste

Table: Fission probabilities of selected actinides, thermal vs. fast neutrons.

Isotope	Thermal Fission Cross Section	Thermal Fission %	Fast Fission Cross Section	Fast Fission %
Th-232	nil	1 (non-fissile)	0.350 barn	3 (non-fissile)
U-232	76.66 barn	59	2.370 barn	95
U-233	531.2 barn	89	2.450 barn	93
U-235	584.4 barn	81	2.056 barn	80
U-238	11.77 microbarn	1 (non-fissile)	1.136 barn	11
Np-237	0.02249 barn	3 (non-fissile)	2.247 barn	27
Pu-238	17.89 barn	7	2.721 barn	70
Pu-239	747.4 barn	63	2.338 barn	85
Pu-240	58.77 barn	1 (non-fissile)	2.253 barn	55
Pu-241	1012 barn	75	2.298 barn	87
Pu-242	0.002557 barn	1 (non-fissile)	2.027 barn	53
Am-241	600.4 barn	1 (non-fissile)	0.2299 microbarn	21
Am-242m	6409 barn	75	2.550 barn	94
Am-243	0.1161 barn	1 (non-fissile)	2.140 barn	23
Cm-242	5.064 barn	1 (non-fissile)	2.907 barn	10
Cm-243	617.4 barn	78	2.500 barn	94
Cm-244	1.037 barn	4 (non-fissile)	0.08255 microbarn	33

Breeder reactors could, in principle, extract almost all of the energy contained in uranium or thorium, decreasing fuel requirements by a factor of 100 compared to widely used once-through light water reactors, which extract less than 1% of the energy in the uranium mined from the earth. The high fuel-efficiency of breeder reactors could greatly reduce concerns about fuel supply or energy used in mining. Adherents claim that with seawater uranium extraction, there would be enough fuel for breeder reactors to satisfy our energy needs for 5 billion years at 1983's total energy consumption rate, thus making nuclear energy effectively a renewable energy.

Nuclear waste became a greater concern by the 1990s. In broad terms, spent nuclear fuel has two main components. The first consists of fission products, the leftover fragments of fuel atoms after they have been split to release energy. Fission products come in dozens of elements and hundreds of isotopes, all of them lighter than uranium. The second main component of spent fuel is transuranics (atoms heavier than uranium), which are generated from uranium or heavier atoms in the fuel when they absorb neutrons but do not undergo fission. All transuranic isotopes fall within the actinide series on the periodic table, and so they are frequently referred to as the actinides.

The physical behavior of the fission products is markedly different from that of the transuranics. In particular, fission products do not themselves undergo fission, and therefore cannot be used for nuclear weapons. Furthermore, only seven long-lived fission product isotopes have half-lives longer than a hundred years, which makes their geological storage or disposal less problematic than for transuranic materials.

With increased concerns about nuclear waste, breeding fuel cycles became interesting again because they can reduce actinide wastes, particularly plutonium and minor actinides. Breeder reactors are designed to fission the actinide wastes as fuel, and thus convert them to more fission products.

After "spent nuclear fuel" is removed from a light water reactor, it undergoes a complex decay profile as each nuclide decays at a different rate. Due to a physical oddity , there is a large gap in the decay half-lives of fission products compared to transuranic isotopes. If the transuranics are left in the spent fuel, after 1,000 to 100,000 years, the slow decay of these transuranics would generate most of the radioactivity in that spent fuel. Thus, removing the transuranics from the waste eliminates much of the long-term radioactivity of spent nuclear fuel.

Today's commercial light water reactors do breed some new fissile material, mostly in the form of plutonium. Because commercial reactors were never designed as breeders, they do not convert enough uranium-238 into plutonium to replace the uranium-235 consumed. Nonetheless, at least one-third of the power produced by commercial nuclear reactors comes from fission of plutonium generated within the fuel. Even with this level of plutonium consumption, light water reactors consume only part of the plutonium and minor actinides they produce, and nonfissile isotopes of plutonium build up, along with significant quantities of other minor actinides.

Conversion Ratio, Break-even, Breeding Ratio, Doubling Time and Burnup

One measure of a reactor's performance is the "conversion ratio," defined as the ratio of new fissile atoms produced to fissile atoms consumed. All proposed nuclear reactors except specially designed and operated actinide burners experience some degree of conversion. As long as there is any amount of a fertile material within the neutron flux of the reactor, some new fissile material is always created. When the conversion ratio is greater than 1, it is often called the "breeding ratio".

For example, commonly used light water reactors have a conversion ratio of approximately 0.6. Pressurized heavy water reactors (PHWR) running on natural uranium have a conversion ratio of 0.8. In a breeder reactor, the conversion ratio is higher than 1. "Break-even" is achieved when the conversion ratio reaches 1.0 and the reactor produces as much fissile material as it uses.

The doubling time is the amount of time it would take for a breeder reactor to produce enough new fissile material to replace the original fuel and additionally produce an equivalent amount of fuel for another nuclear reactor. This was considered an important measure of breeder performance in early years, when uranium was thought to be scarce. However, since uranium is more abundant than thought in the early days of nuclear reactor development, and given the amount of plutonium available in spent reactor fuel, doubling time has become a less-important metric in modern breeder-reactor design.

"Burnup" is a measure of how much energy has been extracted from a given mass of heavy metal in fuel, often expressed (for power reactors) in terms of gigawatt-days per ton of heavy metal. Burnup is an important factor in determining the types and abundances of isotopes produced by a fission reactor. Breeder reactors, by design, have extremely high burnup compared to a conventional reactor, as breeder reactors produce much more of their waste in the form of fission products, while most or all of the actinides are meant to be fissioned and destroyed.

In the past, breeder-reactor development focused on reactors with low breeding ratios, from 1.01 for the Shippingport Reactor running on thorium fuel and cooled by conventional light water to over 1.2 for the Soviet BN-350 liquid-metal-cooled reactor. Theoretical models of breeders with liquid sodium coolant flowing through tubes inside fuel elements ("tube-in-shell" construction) suggest breeding ratios of at least 1.8 are possible on an industrial scale. The Soviet BR-1 test reactor achieved a breeding ratio of 2.5 under non-commercial conditions.

Types of Breeder Reactor

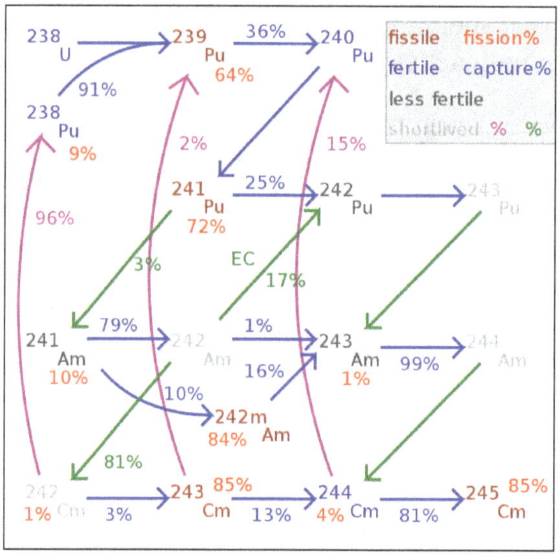

In figure, production of heavy transuranic actinides in current thermal-neutron fission reactors through neutron capture and decays. Starting at uranium-238, isotopes of plutonium, americium, and curium are all produced. In a fast neutron-breeder reactor, all these isotopes may be burned as fuel.

Many types of breeder reactor are possible:

A 'breeder' is simply a reactor designed for very high neutron economy with an associated conversion rate higher than 1.0. In principle, almost any reactor design could be tweaked to become a breeder. An example of this process is the evolution of the Light Water Reactor, a very heavily moderated thermal design, into the Super Fast Reactor concept, using light water in an extremely low-density supercritical form to increase the neutron economy high enough to allow breeding.

Aside from water cooled, there are many other types of breeder reactor currently envisioned as possible. These include molten-salt cooled, gas cooled, and liquid-metal cooled designs in many variations. Almost any of these basic design types may be fueled by uranium, plutonium, many minor actinides, or thorium, and they may be designed for many different goals, such as creating more fissile fuel, long-term steady-state operation, or active burning of nuclear wastes.

Extant reactor designs are sometimes divided into two broad categories based upon their neutron spectrum, which generally separates those designed to use primarily uranium and transuranics from those designed to use thorium and avoid transuranics. These designs are:

- Fast breeder reactor (FBR) which use fast (i.e.: unmoderated) neutrons to breed fissile plutonium and possibly higher transuranics from fertile uranium-238. The fast spectrum is flexible enough that it can also breed fissile uranium-233 from thorium, if desired.

- Thermal breeder reactor which use thermal-spectrum (i.e.: moderated) neutrons to breed fissile uranium-233 from thorium (thorium fuel cycle). Due to the behavior of the various nuclear fuels, a thermal breeder is thought commercially feasible only with thorium fuel, which avoids the buildup of the heavier transuranics.

Reprocessing

Fission of the nuclear fuel in any reactor produces neutron-absorbing fission products. Because of this unavoidable physical process, it is necessary to reprocess the fertile material from a breeder reactor to remove those neutron poisons. This step is required to fully utilize the ability to breed as much or more fuel than is consumed. All reprocessing can present a proliferation concern, since it extracts weapons-usable material from spent fuel. The most-common reprocessing technique, PUREX, presents a particular concern, since it was expressly designed to separate pure plutonium. Early proposals for the breeder-reactor fuel cycle posed an even greater proliferation concern because they would use PUREX to separate plutonium in a highly attractive isotopic form for use in nuclear weapons.

Several countries are developing reprocessing methods that do not separate the plutonium from the other actinides. For instance, the non-water-based pyrometallurgical electrowinning process, when used to reprocess fuel from an integral fast reactor, leaves large amounts of radioactive actinides in the reactor fuel. More-conventional water-based reprocessing systems include SANEX, UNEX, DIAMEX, COEX, and TRUEX, and proposals to combine PUREX with co-processes.

All these systems have modestly better proliferation resistance than PUREX, though their adoption rate is low.

In the thorium cycle, thorium-232 breeds by converting first to protactinium-233, which then decays to uranium-233. If the protactinium remains in the reactor, small amounts of uranium-232 are also produced, which has the strong gamma emitter thallium-208 in its decay chain. Similar to uranium-fueled designs, the longer the fuel and fertile material remain in the reactor, the more of these undesirable elements build up. In the envisioned commercial thorium reactors, high levels of uranium-232 would be allowed to accumulate, leading to extremely high gamma-radiation doses from any uranium derived from thorium. These gamma rays complicate the safe handling of a weapon and the design of its electronics; this explains why uranium-233 has never been pursued for weapons beyond proof-of-concept demonstrations.

While the thorium cycle may be proliferation-resistant with regard to uranium-233 extraction from fuel (because of the presence of uranium-232), it poses a proliferation risk from an alternate route of uranium-233 extraction, which involves chemically extracting protactinium-233 and allowing it to decay to pure uranium-233 outside of the reactor. This process could happen beyond the oversight of organizations such as the International Atomic Energy Agency (IAEA).

Waste Reduction

Actinides and fission products by half-life								
Actinides by decay chain				Half-life range (y)	Fission products of ^{235}U by yield			
4n	4n+1	4n+2	4n+3			4.5–7%	0.04–1.25%	<0.001%
^{228}Ra№				4–6	†		^{155}Euþ	
^{244}Cm^{f}	^{241}Pu^{f}	^{250}Cf	^{227}Ac№	10–29		^{90}Sr	^{85}Kr	^{113m}Cdþ
^{232}U^{f}		^{238}Pu^{f}	^{243}Cm^{f}	29–97		^{137}Cs	^{151}Smþ	^{121m}Sn
^{248}Bk	^{249}Cf^{f}	^{242m}Am^{f}		141–351	No fission products have a half-life in the range of 100–210 k years ...			
	^{241}Am^{f}		^{251}Cf^{f}	430–900				
		^{226}Ra№	^{247}Bk	1.3 k – 1.6 k				
^{240}Pu	^{229}Th	^{246}Cm^{f}	^{243}Am^{f}	4.7 k – 7.4 k				
	^{245}Cm^{f}	^{250}Cm		8.3 k – 8.5 k				
			^{239}Pu^{f}	24.1 k				
		^{230}Th№	^{231}Pa№	32 k – 76 k				
^{236}Np^{f}	^{233}U^{f}	^{234}U№		150 k – 250 k	‡	^{99}Tc₡	^{126}Sn	
^{248}Cm		^{242}Pu		327 k – 375 k			^{79}Se₡	
				1.53 M		^{93}Zr		
	^{237}Np^{f}			2.1 M – 6.5 M		^{135}Cs₡	^{107}Pd	
^{236}U			^{247}Cm^{f}	15 M – 24 M			^{129}I₡	
^{244}Pu№				80 M	... nor beyond 15.7 M years			
^{232}Th№		^{238}U№	^{235}U^{f}№	0.7 G – 14.1 G				

Legend for superscript symbols. ₡ has thermal neutron capture cross section in the range of 8–50 barns. *f* fissile m metastable isomer. № primarily a naturally occurring radioactive material (NORM). þ neutron poison (thermal neutron capture cross section greater than 3k barns). † range 4–97 y: Medium-lived fission product. ‡ over 200,000 y: Long-lived fission product.

Nuclear waste became a greater concern by the 1990s. Breeding fuel cycles attracted renewed interest because of their potential to reduce actinide wastes, particularly plutonium and minor actinides. Since breeder reactors on a closed fuel cycle would use nearly all of the actinides fed into them as fuel, their fuel requirements would be reduced by a factor of about 100. The volume of waste they generate would be reduced by a factor of about 100 as well. While there is a huge reduction in the *volume* of waste from a breeder reactor, the *activity* of the waste is about the same as that produced by a light-water reactor.

In addition, the waste from a breeder reactor has a different decay behavior, because it is made up of different materials. Breeder reactor waste is mostly fission products, while light-water reactor waste has a large quantity of transuranics. After spent nuclear fuel has been removed from a light-water reactor for longer than 100,000 years, these transuranics would be the main source of radioactivity. Eliminating them would eliminate much of the long-term radioactivity from the spent fuel.

In principle, breeder fuel cycles can recycle and consume all actinides, leaving only fission products. As the graphic in this topic indicates, fission products have a peculiar 'gap' in their aggregate half-lives, such that no fission products have a half-life between 91 years and two hundred thousand years. As a result of this physical oddity, after several hundred years in storage, the activity of the radioactive waste from a Fast Breeder Reactor would quickly drop to the low level of the long-lived fission products. However, to obtain this benefit requires the highly efficient separation of transuranics from spent fuel. If the fuel reprocessing methods used leave a large fraction of the transuranics in the final waste stream, this advantage would be greatly reduced.

Both types of breeding cycles can reduce actinide wastes:

- The fast breeder reactor's fast neutrons can fission actinide nuclei with even numbers of both protons and neutrons. Such nuclei usually lack the low-speed "thermal neutron" resonances of fissile fuels used in LWRs.
- The thorium fuel cycle inherently produces lower levels of heavy actinides. The fertile material in the thorium fuel cycle has an atomic weight of 232, while the fertile material in the uranium fuel cycle has an atomic weight of 238. That mass difference means that thorium-232 requires six more neutron capture events per nucleus before the transuranic elements can be produced. In addition to this simple mass difference, the reactor gets two chances to fission the nuclei as the mass increases: First as the effective fuel nuclei U233, and as it absorbs two more neutrons, again as the fuel nuclei U235.

A reactor whose main purpose is to destroy actinides, rather than increasing fissile fuel-stocks, is sometimes known as a burner reactor. Both breeding and burning depend on good neutron economy, and many designs can do either. Breeding designs surround the core by a breeding blanket of fertile material. Waste burners surround the core with non-fertile wastes to be destroyed. Some designs add neutron reflectors or absorbers.

Breeder Reactor Concepts

There are several concepts for breeder reactors; the two main ones are:

- Reactors with a fast neutron spectrum are called fast breeder reactors (FBR) – these typically utilize uranium-238 as fuel.
- Reactors with a thermal neutron spectrum are called thermal breeder reactors – these typically utilize thorium-232 as fuel.

Fast Breeder Reactor

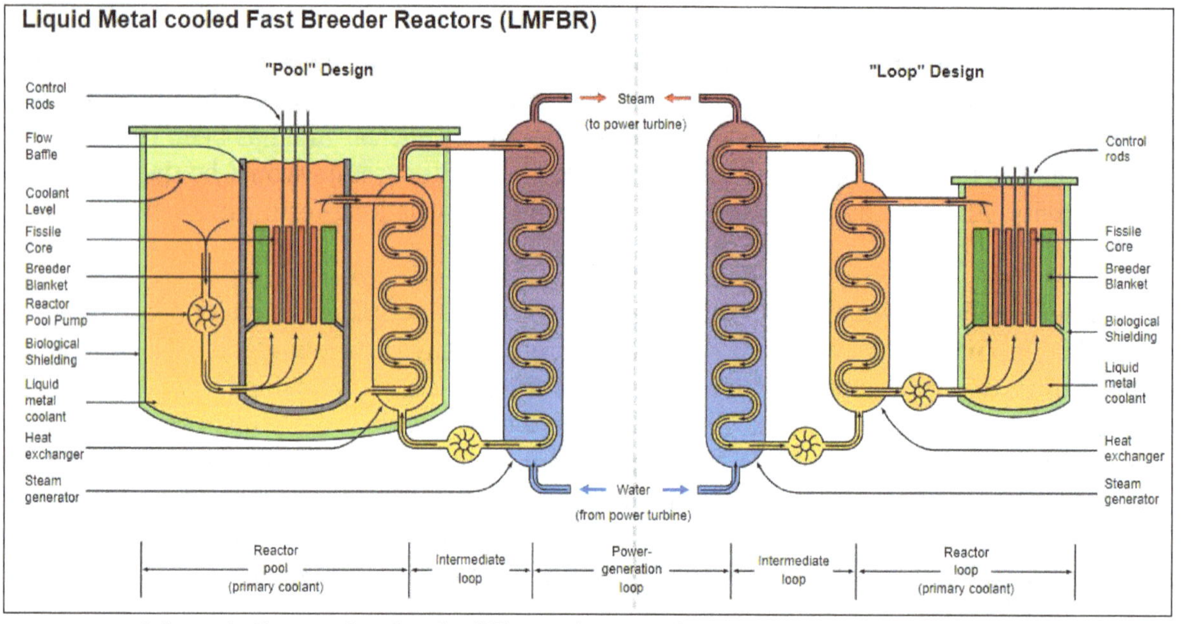

Schematic diagram showing the difference between the Loop and Pool types of LMFBR.

In 2006 all large-scale fast breeder reactor (FBR) power stations were liquid metal fast breeder reactors (LMFBR) cooled by liquid sodium. These have been of one of two designs:

- Loop type, in which the primary coolant is circulated through primary heat exchangers outside the reactor tank (but inside the biological shield due to radioactive sodium-24 in the primary coolant).
- Pool type, in which the primary heat exchangers and pumps are immersed in the reactor tank.

Experimental Breeder Reactor II, which served as the prototype for the Integral Fast Reactor.

All current fast neutron reactor designs use liquid metal as the primary coolant, to transfer heat from the core to steam used to power the electricity generating turbines. FBRs have been built cooled by liquid metals other than sodium—some early FBRs used mercury, other experimental reactors have used a sodium-potassium alloy called NaK. Both have the advantage that they are liquids at room temperature, which is convenient for experimental rigs but less important for pilot or full-scale power stations. Lead and lead-bismuth alloy have also been used.

Four of the proposed generation four reactor types are FBRs:

- Gas-cooled fast reactor (GFR) cooled by helium.
- Sodium-cooled fast reactor (SFR) based on the existing liquid-metal FBR (LMFBR) and integral fast reactor designs.
- Lead-cooled fast reactor (LFR) based on Soviet naval propulsion units.
- Supercritical water reactor (SCWR) based on existing LWR and supercritical boiler technology.

FBRs usually use a mixed oxide fuel core of up to 20% plutonium dioxide (PuO_2) and at least 80% uranium dioxide (UO_2). Another fuel option is metal alloys, typically a blend of uranium, plutonium, and zirconium (used because it is "transparent" to neutrons). Enriched uranium can also be used on its own.

Many designs surround the core in a blanket of tubes that contain non-fissile uranium-238, which, by capturing fast neutrons from the reaction in the core, converts to fissile plutonium-239 (as is some of the uranium in the core), which is then reprocessed and used as nuclear fuel. Other FBR designs rely on the geometry of the fuel itself (which also contains uranium-238), arranged to attain sufficient fast neutron capture. The plutonium-239 (or the fissile uranium-235) fission cross-section is much smaller in a fast spectrum than in a thermal spectrum, as is the ratio between the $^{239}Pu/^{235}U$ fission cross-section and the ^{238}U absorption cross-section. This increases the concentration of $^{239}Pu/^{235}U$ needed to sustain a chain reaction, as well as the ratio of breeding to fission. On the other hand, a fast reactor needs no moderator to slow down the neutrons at all, taking advantage of the fast neutrons producing a greater number of neutrons per fission than slow neutrons. For this reason ordinary liquid water, being a moderator and neutron absorber, is an undesirable primary coolant for fast reactors. Because large amounts of water in the core are required to cool the reactor, the yield of neutrons and therefore breeding of ^{239}Pu are strongly affected. Theoretical work has been done on reduced moderation water reactors, which may have a sufficiently fast spectrum to provide a breeding ratio slightly over 1. This would likely result in an unacceptable power derating and high costs in a liquid-water-cooled reactor, but the supercritical water coolant of the supercritical water reactor (SCWR) has sufficient heat capacity to allow adequate cooling with less water, making a fast-spectrum water-cooled reactor a practical possibility.

The type of coolants, temperatures and fast neutron spectrum puts the fuel cladding material (normally austenitic stainless or ferritic-martensitic steels) under extreme conditions. The understanding of the radiation damage, coolant interactions, stresses and temperatures are necessary for the safe operation of any reactor core. All materials used to date in sodium-cooled fast reactors have known limits, as explored in ONR-RRR-088 review. Oxide Dispersion Strengthened (ODS) steel is viewed as the long-term radiation resistant fuel-cladding material that overcome the shortcomings of today's material choices.

There are only two commercially operating breeder reactors as of 2017: the BN-600 reactor, at 560 MWe, and the BN-800 reactor, at 880 MWe. Both are Russian sodium-cooled reactors.

Integral Fast Reactor

One design of fast neutron reactor, specifically conceived to address the waste disposal and plutonium issues, was the integral fast reactor (IFR, also known as an integral fast breeder reactor, although the original reactor was designed to not breed a net surplus of fissile material).

To solve the waste disposal problem, the IFR had an on-site electrowinning fuel-reprocessing unit that recycled the uranium and all the transuranics (not just plutonium) via electroplating, leaving just short half-life fission products in the waste. Some of these fission products could later be separated for industrial or medical uses and the rest sent to a waste repository. The IFR pyroprocessing system uses molten cadmium cathodes and electrorefiners to reprocess metallic fuel directly on-site at the reactor. Such systems not only co-mingle all the minor actinides with both uranium and plutonium, they are compact and self-contained, so that no plutonium-containing material needs to be transported away from the site of the breeder reactor. Breeder reactors incorporating such technology would most likely be designed with breeding ratios very close to 1.00, so that after an initial loading of enriched uranium and plutonium fuel, the reactor would then be refueled only with small deliveries of natural uranium metal. A quantity of natural uranium metal equivalent to a block about the size of a milk crate delivered once per month would be all the fuel such a 1 gigawatt reactor would need. Such self-contained breeders are currently envisioned as the final self-contained and self-supporting ultimate goal of nuclear reactor designers. The project was canceled in 1994 by United States Secretary of Energy Hazel O'Leary.

Other Fast Reactors

The graphite core of the Molten Salt Reactor Experiment.

Another proposed fast reactor is a fast molten salt reactor, in which the molten salt's moderating properties are insignificant. This is typically achieved by replacing the light metal fluorides (e.g. LiF, BeF_2) in the salt carrier with heavier metal chlorides (e.g., KCl, RbCl, $ZrCl_4$).

Several prototype FBRs have been built, ranging in electrical output from a few light bulbs' equivalent (EBR-I, 1951) to over 1,000 MWe. As of 2006, the technology is not economically competitive to thermal reactor technology, but India, Japan, China, South Korea and Russia are all committing substantial research funds to further development of fast breeder reactors, anticipating that rising uranium prices will change this in the long term. Germany, in contrast, abandoned the technology due to safety concerns. The SNR-300 fast breeder reactor was finished after 19 years despite cost overruns summing up to a total of €3.6 billion, only to then be abandoned.

India is also developing FBR technology using both uranium and thorium feedstocks.

Thermal Breeder Reactor

The Shippingport Reactor, used as a prototype light water breeder for five years beginning in August 1977.

The advanced heavy water reactor (AHWR) is one of the few proposed large-scale uses of thorium. India is developing this technology, motivated by substantial thorium reserves; almost a third of the world's thorium reserves are in India, which lacks significant uranium reserves.

The third and final core of the Shippingport Atomic Power Station 60 MWe reactor was a light water thorium breeder, which began operating in 1977. It used pellets made of thorium dioxide and uranium-233 oxide; initially, the U-233 content of the pellets was 5–6% in the seed region, 1.5–3% in the blanket region and none in the reflector region. It operated at 236 MWt, generating 60 MWe and ultimately produced over 2.1 billion kilowatt hours of electricity. After five years, the core was removed and found to contain nearly 1.4% more fissile material than when it was installed, demonstrating that breeding from thorium had occurred.

The liquid fluoride thorium reactor (LFTR) is also planned as a thorium thermal breeder. Liquid-fluoride reactors may have attractive features, such as inherent safety, no need to manufacture fuel rods and possibly simpler reprocessing of the liquid fuel. This concept was first investigated at the Oak Ridge National Laboratory Molten-Salt Reactor Experiment in the 1960s. From 2012 it became the subject of renewed interest worldwide. Japan, India, China, the UK, as well as private US, Czech and Australian companies have expressed intent to develop and commercialize the technology.

Steam Generators

Steam generators are heat exchangers used to convert water into steam from heat produced in a nuclear reactor core. They are used in pressurized water reactors (PWR) between the primary and secondary coolant loops.

In typical PWR designs, the primary coolant is high-purity water, kept under high pressure so it cannot boil. This primary coolant is pumped through the reactor core where it absorbs heat from the fuel rods. It then passes through the steam generator, where it transfers its heat (via conduction through metal) to lower-pressure water which is allowed to boil.

Unlike PWRs, Boiling water reactors (BWR) do not use steam generators. The primary coolant is allowed to boil directly in the reactor core, and the steam is simply passed through a steam turbine. While theoretically simple, this has a downside for maintenance. While passing through the core, primary coolant water is subjected to high neutron flux. This activates oxygen and dissolved nitrogen in the water. The major reaction is: an atom of oxygen-16 absorbs 1 neutron and emits 1 proton, becoming nitrogen-16. Nitrogen-16 has a 7-second half-life and produces a gamma ray when it decays back to oxygen-16. The 7-second half-life is long enough for the water to circulate out of the reactor. In a BWR, this means that the water may be in the steam turbine when it releases its gamma rays. Although no long-lived radioisotopes are produced by this reaction, the gamma radiation means that humans cannot be present in a BWR's turbine hall during reactor operation and for a short time afterwards.

By contrast, in a PWR, the steam generator separates the activated primary coolant water from the secondary coolant which passes through the steam turbine. Thus, humans can freely access a PWR's turbines and other steam plant components during operation. This reduces maintenance cost and improves up-time.

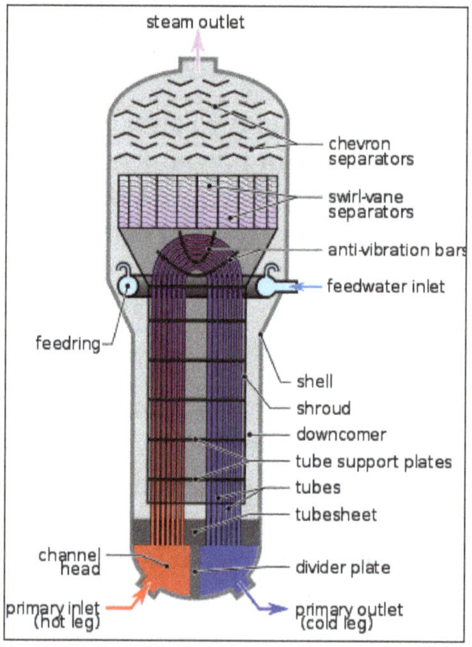

Vertical recirculating-type steam generator (typical of Westinghouse and Combustion Engineering-designed reactors) and components.

In commercial power plants, there are two to four steam generators per reactor; each steam generator can measure up to 70 feet (21 m) in height and weigh as much as 800 tons. Each steam generator can contain anywhere from 3,000 to 16,000 tubes, each about .75 inches (19 mm) in diameter. The coolant (treated water), which is maintained at high pressure to prevent boiling, is pumped through the nuclear reactor core. Heat transfer takes place between the reactor core and the circulating water and the coolant is then pumped through the primary tube side of the steam generator by coolant pumps before returning to the reactor core. This is referred to as the primary loop.

That water flowing through the steam generator boils water on the shell side (which is kept at a lower pressure than the primary side) to produce steam. This is referred to as the secondary loop. The secondary-side steam is delivered to the turbines to make electricity. The steam is subsequently condensed via cooled water from a tertiary loop and returned to the steam generator to be heated once again. The tertiary cooling water may be recirculated to cooling towers where it sheds waste heat before returning to condense more steam. Once-through tertiary cooling may otherwise be provided by a river, lake, or ocean. This primary, secondary, tertiary cooling scheme is the basis of the pressurized water reactor, which is the most common nuclear power plant design worldwide.

In other types of reactors, such as the pressurised heavy water reactors of the CANDU design, the primary fluid is heavy water. Liquid metal cooled reactors such as the Russian BN-600 reactor use a liquid metal, such as sodium, as the primary coolant. These also use heat exchangers between primary metal coolant and the secondary water coolant, and thus their secondary and tertiary cooling is similar to a PWR.

A steam generator's heat-exchange tubes have an important safety role, because they separate radioactive and non-radioactive fluid systems. (The primary coolant becomes briefly radioactive from its exposure to the core, and also has trace amounts of longer-lived radioactive isotopes dissolved in it, such as dissolved atoms of iron from pipes.) Because the primary coolant is at higher pressure, a ruptured heat-exchange tube would cause primary coolant to leak into the secondary loop. Typically this would require the plant to shutdown for repair. To avoid such primary-secondary leaks, steam generator tubes are periodically inspected by eddy-current testing, and individual tubes can be plugged to remove them from operation. As with many nuclear components, mechanical engineers determine the inspection frequency using the known rates of corrosion and crack propagation in the material. If an inspection finds that a tube wall is thin enough that it might corrode through before the next scheduled inspection, the tube is plugged. (Plugging a tube is typically easier than attempting to repair it. There are many small heat-exchange tubes, and steam generators are designed with excess tubes to allow some to be plugged).

Entire steam generators are often replaced in plant mid-life, which is a major undertaking. Most U.S. PWR plants have had steam generators replaced.

Types

Westinghouse and Combustion Engineering designs have vertical U-tubes with inverted tubes for the primary water. Canadian, Japanese, French, and German PWR suppliers use the vertical configuration as well. Russian VVER reactor designs use horizontal steam generators, which have the tubes mounted horizontally. Babcock & Wilcox plants (e.g., Three Mile Island) have smaller steam generators that force water through the top of the OTSGs (once-through steam generators; counter-flow to the feedwater) and out the bottom to be recirculated by the reactor coolant pumps. The horizontal design has proven to be less susceptible to degradation than the vertical U-tube design.

Materials and Construction

The materials that make up the turbine and pipes of a nuclear powered steam generator are specially made and specifically designed to withstand the heat and radiation of the reactor. The water tubes also have to be able to resist corrosion from water for an extended period of time. The pipes that are used in American reactors are made of Inconel, either Alloy 600 or Alloy 690. Alloy 690 is made with extra chromium and most facilities heat treat the metal to make it better able to resist heat and corrosion. The high nickel content in Alloy 600 and Alloy 690 make them well suited for resisting acids and high degrees of stress and temperature.

Degradation

The annealed, or heat treated, Alloy 600 was prone to tube denting and thinning due to water chemistry. Plants that used the Alloy 600 in their water tubes therefore had to install new water chemistry controllers and change the chemicals they put in the water. Due to this, pipe thinning has been taken care of, but on rare occasions, tube denting still occurs, causing leaks and ruptures. The only way to prevent this is regular maintenance and check-ups, but this forces the reactor to shut down. In some cases, plants replaced their Alloy 600 tubes with Alloy 690 tubes and a few plants were shut down. To prevent future problems, manufacturers of steam turbines for nuclear power plants have improved their fabrication techniques and used other materials, such as stainless steel, to prevent tube denting.

Typical Operating Conditions

Steam generators in a "typical" PWR in the USA have the following operating conditions:

Side	Pressure (absolute)	Inlet temperature	Outlet temperature
Primary side (tube side)	15.5 MPa (2,250 psi)	315 °C (599 °F) (liquid water)	275 °C (527 °F) (liquid water)
Secondary side (shell side)	6.2 MPa (900 psi)	220 °C (428 °F) (liquid water)	275 °C (527 °F) (saturated steam)

Tube Material

Various high-performance alloys and superalloys have been used for steam generator tubing, including type 316 stainless steel, Alloy 400, Alloy 600MA (mill annealed), Alloy 600TT (thermally treated), Alloy 690TT, and Alloy 800Mod.

Reactor Coolant Pump

Nuclear power plants rely on cooling systems to ensure safe, continuous operation of the nuclear reactor. Cooling systems naturally ensure a heat transfer from a reactor core to steam generators, which is the main purpose of the cooling systems. Because of the large amount of heat generated in the reactor core by the fission reaction, the cooling systems demand a large volumetric flow of water (~80000 m^3/hr) to ensure a sufficient and safe heat transfer. The cooling water is usually supplied by two or more large centrifugal pumps called reactor coolant pumps (RCPs). RCPs are not usually "safety system", as defined. After the loss of RCPs the reactor must be shutdown immediately. Sufficient and safe residual heat removal is then ensured by a natural circulation flow through the reactor. However, natural circulation is not sufficient to remove the heat being generatcd whcn the reactor is at power.

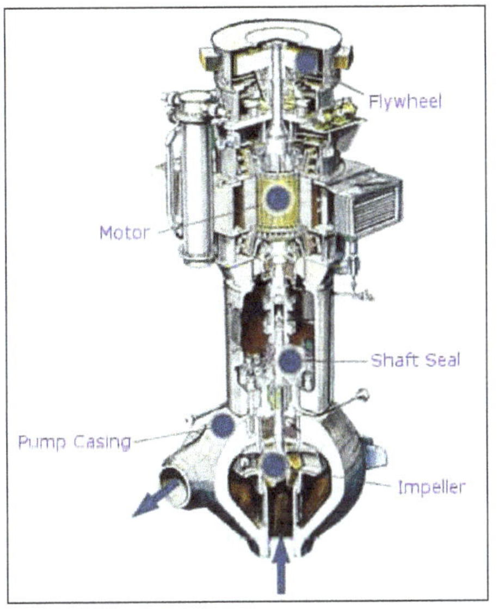

Reactor coolant pumps (RCPs) are used to pump primary coolant around the primary circuit. The purpose of the reactor coolant pump is to provide forced primary coolant flow to remove and transfer the amount of heat generated in the reactor core.There are many designs of these pumps and there are many designs of primary coolant loops. There are significant differences between pumps for different reactor types. Most of PWRs use four RCPs in two or four loops design.

Generally reactor coolant pumps are powerful, they can consume up to 6 MW each and therefore they can be used for heating the primary coolant before a reactor startup.

Most of RCPs are vertical installed on a cold leg of a primary loop, but also a direct connection to a steam generator is possible. The reactor coolant enters the suction side of the pump at high pressure and temperature (~16MPa; 290 °C; 554 °F). The water is increased in velocity by the pump impeller. This increase in velocity is converted to pressure in the discharge volute. At the discharge of the reactor coolant pump, the reactor coolant pressure will be approximately 0, 5 MPa higher than the inlet pressure. After the coolant leaves the discharge side of the pump, it will enter the cold leg and continue to the reactor. The coolant will then pass through the nuclear core and through the fuel, where collects heat and is sent back to the steam generators.

The Main Components of a Reactor Coolant Pump

- Electric motor: The motor is a large, air or water (seal-less RCPs) cooled, induction motor.
- Impeller: Impeller is a rotor used to increase the pressure and flow of a coolant.
- Shaft (Rotor): Shaft is a mechanical component for transmitting torque from the motor to the impeller.
- Shaft seal package: Shaft seal package is used to prevent any water from leaking up the shaft into the containment.
- Bearings: Bearings constrain relative motion of the shaft (rotor) and reduce friction between the rotating shaft and the stator. RCPs usually use a combination of fluid dynamic bearings and hydrostatic bearings in the radial bearing assembly (water lubricated; close to the primary coolant) and oil lubricated bearings used in the thrust (axial) bearing assembly Flywheel. The flywheel provides flow coastdown in case of loss of power.
- Auxilliary systems: Oil lubrication system, oil lift system, seal leakoff system, seal cooling system etc.

Use of Shaft Seals

The seal package is located on the shaft between the electric motor and the impeller and prevents any primary coolant from leaking up the shaft into the containment. Any coolant that does leak up the shaft is collected and routed to the seal leakoff system.

- RCPs with shaft seals: In order to maintain pump pressure and restrict water volume loss, the pumps typically utilize a multi-stage mechanical face seal system. This configuration allows use of flywheel, which provides rotating inertia to ensure a slow decrease in coolant flow in order to prevent fuel damage as a result of a loss of power to the pump motors. An integrity of the pressure boundary in the event of a postulated flywheel failure has to be proved.
- Seal-less RCPs: This pumps do not have any shaft seals and any large flywheels. Entire RCP systems (motor, impeller, shaft, fluid bearings) are sealed at high pressure side of primary circuit. Such RCPs (Canned motor pumps) do not require so much external systems (no lube oil system, seal injection and leak-off system) as the RCPs with shaft seals. These pumps can be more reliable, but sufficient rotating inertia (internal flywheels) to provide flow coastdown has to be proved.

Cooling Tower

A cooling tower is a heat rejection device that rejects waste heat to the atmosphere through the cooling of a water stream to a lower temperature. Cooling towers may either use the evaporation of water to remove process heat and cool the working fluid to near the wet-bulb air temperature or, in the case of *closed circuit dry cooling towers*, rely solely on air to cool the working fluid to near the dry-bulb air temperature.

Common applications include cooling the circulating water used in oil refineries, petrochemical and other chemical plants, thermal power stations, nuclear power stations and HVAC systems for cooling buildings. The classification is based on the type of air induction into the tower: the main types of cooling towers are natural draft and induced draft cooling towers.

Cooling towers vary in size from small roof-top units to very large hyperboloid structures that can be up to 200 metres (660 ft) tall and 100 metres (330 ft) in diameter, or rectangular structures that can be over 40 metres (130 ft) tall and 80 metres (260 ft) long. The hyperboloid cooling towers are often associated with nuclear power plants, although they are also used in some coal-fired plants and to some extent in some large chemical and other industrial plants. Although these large towers are very prominent, the vast majority of cooling towers are much smaller, including many units installed on or near buildings to discharge heat from air conditioning.

Classification by Use

Heating, Ventilation and Air Conditioning (HVAC)

Two HVAC cooling towers on the rooftop of a shopping center (Germany).

An HVAC (heating, ventilating, and air conditioning) cooling tower is used to dispose of ("reject") unwanted heat from a chiller. Water-cooled chillers are normally more energy efficient than air-cooled chillers due to heat rejection to tower water at or near wet-bulb temperatures. Air-cooled chillers must reject heat at the higher dry-bulb temperature, and thus have a lower average reverse-Carnot cycle effectiveness. In areas with a hot climate, large office buildings, hospitals, and schools typically use one or more cooling towers as part of their air conditioning systems. Generally, industrial cooling towers are much larger than HVAC towers.

Cell of an open loop cooling tower with fill material, and circulating water visible.

HVAC use of a cooling tower pairs the cooling tower with a water-cooled chiller or water-cooled condenser. A *ton* of air-conditioning is defined as the removal of 12,000 BTU/hour (3500 W). The *equivalent ton* on the cooling tower side actually rejects about 15,000 BTU/hour (4400 W) due to the additional waste heat-equivalent of the energy needed to drive the chiller's compressor. This *equivalent ton* is defined as the heat rejection in cooling 3 US gallons/minute (1,500 pound/hour) of water 10 °F (6 °C), which amounts to 15,000 BTU/hour, assuming a chiller coefficient of performance (COP) of 4.0. This COP is equivalent to an energy efficiency ratio (EER) of 14.

Cooling towers are also used in HVAC systems that have multiple water source heat pumps that share a common piping *water loop*. In this type of system, the water circulating inside the water loop removes heat from the condenser of the heat pumps whenever the heat pumps are working in the cooling mode, then the externally mounted cooling tower is used to remove heat from the water loop and reject it to the atmosphere. By contrast, when the heat pumps are working in heating mode, the condensers draw heat out of the loop water and reject it into the space to be heated. When the water loop is being used primarily to supply heat to the building, the cooling tower is normally shut down (and may be drained or winterized to prevent freeze damage), and heat is supplied by other means, usually from separate boilers.

Industrial Cooling Towers

Industrial cooling towers for a power plant.

Industrial Cooling Towers for Fruit Processing Industry.

Industrial cooling towers can be used to remove heat from various sources such as machinery or heated process material. The primary use of large, industrial cooling towers is to remove the heat absorbed in the circulating cooling water systems used in power plants, petroleum refineries, petrochemical plants, natural gas processing plants, food processing plants, semi-conductor plants, and for other industrial facilities such as in condensers of distillation columns, for cooling liquid in crystallization, etc. The circulation rate of cooling water in a typical 700 MW coal-fired power plant with a cooling tower amounts to about 71,600 cubic metres an hour (315,000 US gallons per minute) and the circulating water requires a supply water make-up rate of perhaps 5 percent (i.e., 3,600 cubic metres an hour).

If that same plant had no cooling tower and used once-through cooling water, it would require about 100,000 cubic metres an hour A large cooling water intake typically kills millions of fish and larvae annually, as the organisms are impinged on the intake screens. A large amount of water would have to be continuously returned to the ocean, lake or river from which it was obtained and continuously re-supplied to the plant. Furthermore, discharging large amounts of hot water may raise the temperature of the receiving river or lake to an unacceptable level for the local ecosystem. Elevated water temperatures can kill fish and other aquatic organisms, or can also cause an increase in undesirable organisms such as invasive species of zebra mussels or algae. A cooling tower serves to dissipate the heat into the atmosphere instead and wind and air diffusion spreads the heat over a much larger area than hot water can distribute heat in a body of water. Evaporative cooling water cannot be used for subsequent purposes (other than rain somewhere), whereas surface-only cooling water can be re-used. Some coal-fired and nuclear power plants located in coastal areas do make use of once-through ocean water. But even there, the offshore discharge water outlet requires very careful design to avoid environmental problems.

Petroleum refineries also have very large cooling tower systems. A typical large refinery processing 40,000 metric tonnes of crude oil per day (300,000 barrels (48,000 m^3) per day) circulates about 80,000 cubic metres of water per hour through its cooling tower system.

The world's tallest cooling tower is the 202 metres (663 ft) tall cooling tower of Kalisindh Thermal Power Station in Jhalawar, Rajasthan, India.

Field erected cooling tower.

Classification by Build

Package Type

Field Erected Cooling Towers.

Brotep-Eco cooling tower.

These types of cooling towers are factory preassembled, and can be simply transported on trucks, as they are compact machines. The capacity of package type towers is limited and, for that reason, they are usually preferred by facilities with low heat rejection requirements such as food processing plants, textile plants, some chemical processing plants, or buildings like hospitals, hotels, malls, automotive factories etc.

Due to their frequent use in or near residential areas, sound level control is a relatively more important issue for package type cooling towers.

Package cooling tower.

Field Erected Type

Facilities such as power plants, steel processing plants, petroleum refineries, or petrochemical plants usually install field erected type cooling towers due to their greater capacity for heat rejection. Field erected towers are usually much larger in size compared to the package type cooling towers.

A typical field erected cooling tower has a pultruded fiber-reinforced plastic (FRP) structure, FRP cladding, a mechanical unit for air draft, drift eliminator.

Heat Transfer Methods

With respect to the heat transfer mechanism employed, the main types are:

- Wet cooling towers (or open circuit cooling towers) operate on the principle of evaporative cooling. The working fluid and the evaporated fluid (usually water) are one and the same.
- Closed circuit cooling towers (or fluid coolers) pass the working fluid through a tube bundle, upon which clean water is sprayed and a fan-induced draft applied. The resulting heat transfer performance is close to that of a wet cooling tower, with the advantage of protecting the working fluid from environmental exposure and contamination.
- Dry cooling towers are closed circuit cooling towers which operate by heat transfer through a surface that separates the working fluid from ambient air, such as in a tube to air heat exchanger, utilizing convective heat transfer. They do not use evaporation.

- Hybrid cooling towers are closed circuit cooling towers that can switch between wet and dry operation. This helps balance water and energy savings across a variety of weather conditions.

In a wet cooling tower (or open circuit cooling tower), the warm water can be cooled to a temperature *lower* than the ambient air dry-bulb temperature, if the air is relatively dry . As ambient air is drawn past a flow of water, a small portion of the water evaporates, and the energy required to evaporate that portion of the water is taken from the remaining mass of water, thus reducing its temperature. Approximately 970 BTU of heat energy is absorbed for each pound of evaporated water (2 MJ/kg). Evaporation results in saturated air conditions, lowering the temperature of the water processed by the tower to a value close to wet-bulb temperature, which is lower than the ambient dry-bulb temperature, the difference determined by the initial humidity of the ambient air.

To achieve better performance (more cooling), a medium called *fill* is used to increase the surface area and the time of contact between the air and water flows. *Splash fill* consists of material placed to interrupt the water flow causing splashing. *Film fill* is composed of thin sheets of material (usually PVC) upon which the water flows. Both methods create increased surface area and time of contact between the fluid (water) and the gas (air), to improve heat transfer.

Air Flow Generation Methods

Access stairs at the base of a massive hyperboloid cooling tower give a sense of its scale (UK).

With respect to drawing air through the tower, there are three types of cooling towers:

- Natural draft: Utilizes buoyancy via a tall chimney. Warm, moist air naturally rises due to the density differential compared to the dry, cooler outside air. Warm moist air is less dense than drier air at the same pressure. This moist air buoyancy produces an upwards current of air through the tower.

- Mechanical draft: Uses power-driven fan motors to force or draw air through the tower.

- Induced draft: A mechanical draft tower with a fan at the discharge (at the top) which pulls air up through the tower. The fan *induces* hot moist air out the discharge. This produces low entering and high exiting air velocities, reducing the possibility of *recirculation* in which discharged air flows back into the air intake. This fan/fin arrangement is also known as *draw-through*.

- Forced draft: A mechanical draft tower with a blower type fan at the intake. The fan *forces* air into the tower, creating high entering and low exiting air velocities. The low exiting velocity is much more susceptible to recirculation. With the fan on the air intake, the fan is more susceptible to complications due to freezing conditions. Another disadvantage is that a forced draft design typically requires more motor horsepower than an equivalent induced draft design. The benefit of the forced draft design is its ability to work with high static pressure. Such setups can be installed in more-confined spaces and even in some indoor situations. This fan/fin geometry is also known as *blow-through.*
- Fan assisted natural draft: A hybrid type that appears like a natural draft setup, though airflow is assisted by a fan.

Hyperboloid (sometimes incorrectly known as hyperbolic) cooling towers have become the design standard for all natural-draft cooling towers because of their structural strength and minimum usage of material. The hyperboloid shape also aids in accelerating the upward convective air flow, improving cooling efficiency. These designs are popularly associated with nuclear power plants. However, this association is misleading, as the same kind of cooling towers are often used at large coal-fired power plants as well. Conversely, not all nuclear power plants have cooling towers, and some instead cool their heat exchangers with lake, river or ocean water.

Thermal efficiencies up to 92% have been observed in hybrid cooling towers.

Categorization by Air-to-water Flow

Crossflow

Typically lower initial and long-term cost, mostly due to pump requirements.

Crossflow is a design in which the air flow is directed perpendicular to the water flow. Air flow enters one or more vertical faces of the cooling tower to meet the fill material. Water flows (perpendicular to the air) through the fill by gravity. The air continues through the fill and thus past the water flow into an open plenum volume. Lastly, a fan forces the air out into the atmosphere.

Package crossflow cooling tower.

Mechanical draft crossflow cooling tower used in an HVAC application.

A *distribution* or *hot water basin* consisting of a deep pan with holes or *nozzles* in its bottom is located near the top of a crossflow tower. Gravity distributes the water through the nozzles uniformly across the fill material.

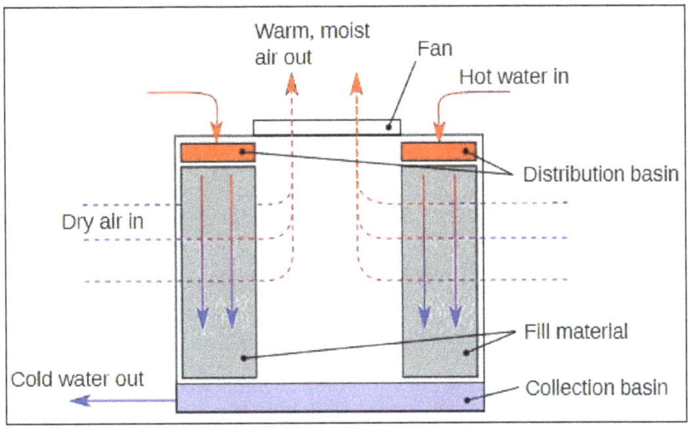

Advantages of the crossflow design:

- Gravity water distribution allows smaller pumps and maintenance while in use.
- Non-pressurized spray simplifies variable flow.

Disadvantages of the crossflow design:

- More prone to freezing than counterflow designs.
- Variable flow is useless in some conditions.
- More prone to dirt buildup in the fill than counterflow designs, especially in dusty or sandy areas.

Counterflow

Induced Draft Counterflow Cooling Towers.

In a counterflow design, the air flow is directly opposite to the water flow. Air flow first enters an open area beneath the fill media, and is then drawn up vertically. The water is sprayed through pressurized nozzles near the top of the tower, and then flows downward through the fill, opposite to the air flow.

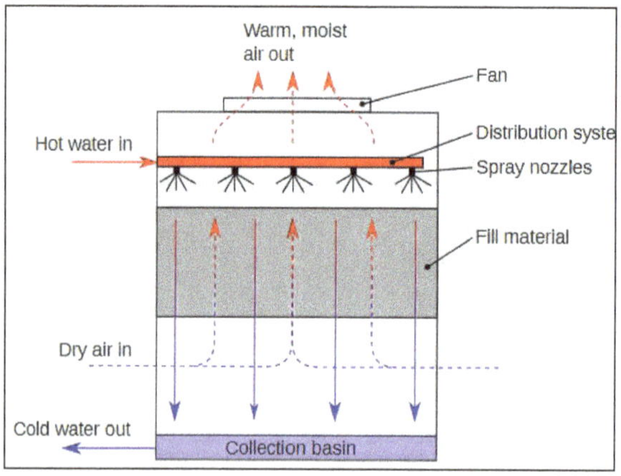

Advantages of the counterflow design:

- Spray water distribution makes the tower more freeze-resistant.
- Breakup of water in spray makes heat transfer more efficient.

Disadvantages of the counterflow design:

- Typically higher initial and long-term cost, primarily due to pump requirements.
- Difficult to use variable water flow, as spray characteristics may be negatively affected.
- Typically noisier, due to the greater water fall height from the bottom of the fill into the cold water basin.

Common Aspects

Common aspects of both designs:

- The interactions of the air and water flow allow a partial equalization of temperature, and evaporation of water.
- The air, now saturated with water vapor, is discharged from the top of the cooling tower.
- A "collection basin" or "cold water basin" is used to collect and contain the cooled water after its interaction with the air flow.

Both crossflow and counterflow designs can be used in natural draft and in mechanical draft cooling towers.

Wet Cooling Tower Material Balance

Quantitatively, the material balance around a wet, evaporative cooling tower system is governed by the operational variables of make-up volumetric flow rate, evaporation and windage losses, draw-off rate, and the concentration cycles.

In the adjacent diagram, water pumped from the tower basin is the cooling water routed through the process coolers and condensers in an industrial facility. The cool water absorbs heat from the hot process streams which need to be cooled or condensed, and the absorbed heat warms the circulating water (C). The warm water returns to the top of the cooling tower and trickles downward over the fill material inside the tower. As it trickles down, it contacts ambient air rising up through the tower either by natural draft or by forced draft using large fans in the tower. That contact causes a small amount of the water to be lost as windage or drift (W) and some of the water (E) to evaporate. The heat required to evaporate the water is derived from the water itself, which cools the water back to the original basin water temperature and the water is then ready to recirculate. The evaporated water leaves its dissolved salts behind in the bulk of the water which has not been evaporated, thus raising the salt concentration in the circulating cooling water. To prevent the salt concentration of the water from becoming too high, a portion of the water is drawn off or blown down (D) for disposal. Fresh water make-up (M) is supplied to the tower basin to compensate for the loss of evaporated water, the windage loss water and the draw-off water.

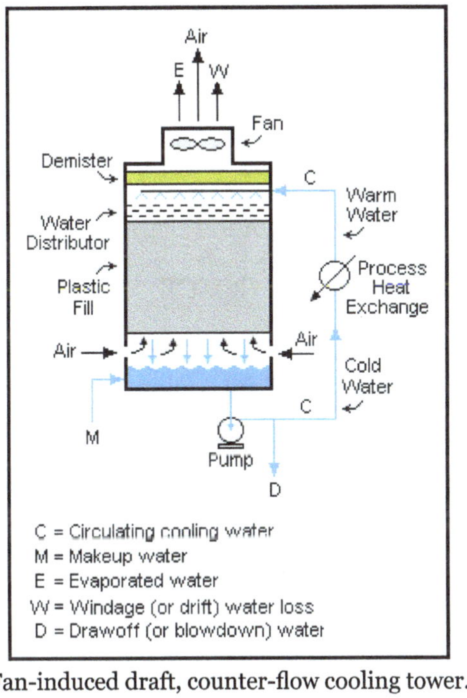

Fan-induced draft, counter-flow cooling tower.

Figure indicate using these flow rates and concentration dimensional units:

M	= Make-up water in m^3/h
C	= Circulating water in m^3/h
D	= Draw-off water in m^3/h
E	= Evaporated water in m^3/h
W	= Windage loss of water in m^3/h
X	= Concentration in ppmw (of any completely soluble salts usually chlorides)
X_M	= Concentration of chlorides in make-up water (M), in ppmw
X_C	= Concentration of chlorides in circulating water (C), in ppmw
Cycles	= Cycles of concentration = X_C / X_M (dimensionless)
ppmw	= parts per million by weight

A water balance around the entire system is then:

$$M = E + D + W$$

Since the evaporated water (E) has no salts, a chloride balance around the system is:

$$MX_M = DX_C + WX_C = X_C(D+W)$$

and, therefore:

$$\frac{X_C}{X_M} = \text{Cycles of concentration} = \frac{M}{(D+W)} = \frac{M}{(M-E)} = 1 + \frac{E}{(D+W)}$$

From a simplified heat balance around the cooling tower:

$$E = \frac{C \Delta T c_p}{H_V}$$

where,

- HV = latent heat of vaporization of water = 2260 kJ / kg.
- ΔT = water temperature difference from tower top to tower bottom, in °C.
- cp = specific heat of water = 4.184 kJ / (kg °C).

Windage (or drift) losses (W) is the amount of total tower water flow that is entrained in the flow of air to the atmosphere. From large-scale industrial cooling towers, in the absence of manufacturer's data, it may be assumed to be:

- W = 0.3 to 1.0 percent of C for a natural draft cooling tower without windage drift eliminators.
- W = 0.1 to 0.3 percent of C for an induced draft cooling tower without windage drift eliminators.
- W = about 0.005 percent of C (or less) if the cooling tower has windage drift eliminators.
- W = about 0.0005 percent of C (or less) if the cooling tower has windage drift eliminators and uses sea water as make-up water.

Cycles of Concentration

Cycle of concentration represents the accumulation of dissolved minerals in the recirculating cooling water. Discharge of draw-off (or blowdown) is used principally to control the buildup of these minerals.

The chemistry of the make-up water, including the amount of dissolved minerals, can vary widely. Make-up waters low in dissolved minerals such as those from surface water supplies (lakes, rivers etc.) tend to be aggressive to metals (corrosive). Make-up waters from ground water supplies (such as wells) are usually higher in minerals, and tend to be scaling (deposit minerals). Increasing the amount of minerals present in the water by cycling can make water less aggressive to piping, however, excessive levels of minerals can cause scaling problems.

As the cycles of concentration increase, the water may not be able to hold the minerals in solution. When the solubility of these minerals have been exceeded they can precipitate out as mineral solids and cause fouling and heat exchange problems in the cooling tower or the heat exchangers. The temperatures of the recirculating water, piping and heat exchange surfaces determine if and where minerals will precipitate from the recirculating water. Often a professional water treatment consultant will evaluate the make-up water and the operating conditions of the cooling tower and recommend an appropriate range for the cycles of concentration. The use of water treatment chemicals, pretreatment such as water softening, pH adjustment, and other techniques can affect the acceptable range of cycles of concentration.

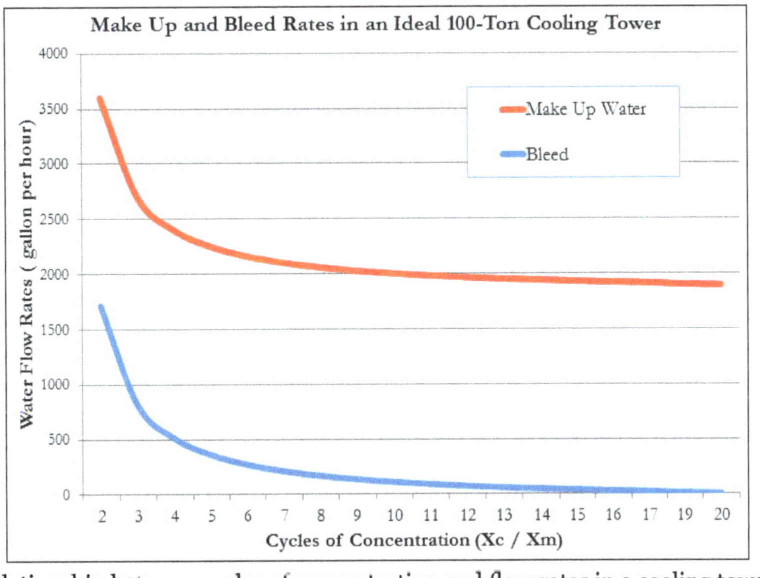

Relationship between cycles of concentration and flow rates in a cooling tower.

Concentration cycles in the majority of cooling towers usually range from 3 to 7. In the United States, many water supplies use well water which has significant levels of dissolved solids. On the other hand, one of the largest water supplies, for New York City, has a surface rainwater source quite low in minerals; thus cooling towers in that city are often allowed to concentrate to 7 or more cycles of concentration.

Since higher cycles of concentration represent less make-up water, water conservation efforts may focus on increasing cycles of concentration. Highly treated recycled water may be an effective means of reducing cooling tower consumption of potable water, in regions where potable water is scarce.

Maintenance

Surfaces with any visible biofilm (i.e., slime) should be cleaned. Disinfectant and other chemical levels in cooling towers and hot tubs should be continuously maintained and regularly monitored. Regular checks of water quality (specifically the aerobic bacteria levels) using dipslides should be taken as the presence of other organisms can support legionella by producing the organic nutrients that it needs to thrive.

Water Treatment

Besides treating the circulating cooling water in large industrial cooling tower systems to minimize scaling and fouling, the water should be filtered to remove particulates, and also be dosed with biocides and algaecides to prevent growths that could interfere with the continuous flow of the water. Under certain conditions, a biofilm of micro-organisms such as bacteria, fungi and algae can grow very rapidly in the cooling water, and can reduce the heat transfer efficiency of the cooling tower. Biofilm can be reduced or prevented by using chlorine or other chemicals. A normal industrial practice is to use two biocides, such as oxidizing and non-oxidizing types to complement each other's strengths and weaknesses, and to ensure a broader spectrum of attack. In most cases, a continual low level oxidizing biocide is used, then alternating to a periodic shock dose of non-oxidizing biocides.

Legionnaires' Disease

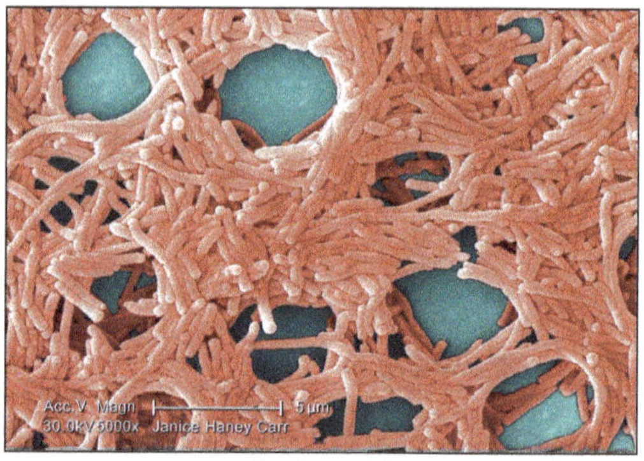

Legionella pneumophila (5000x magnification).

Another very important reason for using biocides in cooling towers is to prevent the growth of *Legionella*, including species that cause legionellosis or Legionnaires' disease, most notably *L. pneumophila*, or *Mycobacterium avium*. The various *Legionella* species are the cause of Legionnaires' disease in humans and transmission is via exposure to aerosols—the inhalation of mist droplets containing the bacteria. Common sources of *Legionella* include cooling towers used in open recirculating evaporative cooling water systems, domestic hot water systems, fountains, and similar disseminators that tap into a public water supply. Natural sources include freshwater ponds and creeks.

A multitude of microscopic organisms such as bacterial colonies, fungi, and algae can easily thrive within the moderately high temperatures present inside a cooling tower.

French researchers found that *Legionella* bacteria travelled up to 6 kilometres (3.7 mi) through the air from a large contaminated cooling tower at a petrochemical plant in Pas-de-Calais, France. That outbreak killed 21 of the 86 people who had a laboratory-confirmed infection.

Drift (or windage) is the term for water droplets of the process flow allowed to escape in the cooling tower discharge. Drift eliminators are used in order to hold drift rates typically to 0.001–0.005% of the circulating flow rate. A typical drift eliminator provides multiple directional changes of airflow to prevent the escape of water droplets. A well-designed and well-fitted drift eliminator can greatly reduce water loss and potential for *Legionella* or water treatment chemical exposure.

The CDC does not recommend that health-care facilities regularly test for the *Legionella pneumophila* bacteria. Scheduled microbiologic monitoring for *Legionella* remains controversial because its presence is not necessarily evidence of a potential for causing disease. The CDC recommends aggressive disinfection measures for cleaning and maintaining devices known to transmit *Legionella*, but does not recommend regularly-scheduled microbiologic assays for the bacteria. However, scheduled monitoring of potable water within a hospital might be considered in certain settings where persons are highly susceptible to illness and mortality from *Legionella* infection (e.g. hematopoietic stem cell transplantation units, or solid organ transplant units). Also, after an outbreak of legionellosis, health officials agree that monitoring is necessary to identify the source and to evaluate the efficacy of biocides or other prevention measures. Studies have found *Legionella* in 40% to 60% of cooling towers.

Fog Production

Fog produced by Eggborough power station.

Under certain ambient conditions, plumes of water vapor can be seen rising out of the discharge from a cooling tower, and can be mistaken as smoke from a fire. If the outdoor air is at or near saturation, and the tower adds more water to the air, saturated air with liquid water droplets can be discharged, which is seen as fog. This phenomenon typically occurs on cool, humid days, but is rare in many climates. Fog and clouds associated with cooling towers can be described as homogenitus, as with other clouds of man-made origin, such as contrails and ship tracks.

This phenomenon can be prevented by decreasing the relative humidity of the saturated discharge air. For that purpose, in hybrid towers, saturated discharge air is mixed with heated low relative humidity air. Some air enters the tower above drift eliminator level, passing through heat exchangers. The relative humidity of the dry air is even more decreased instantly as being heated while entering the tower. The discharged mixture has a relatively lower relative humidity and the fog is invisible.

Salt Emission Pollution

When wet cooling towers with seawater make-up are installed in various industries located in or near coastal areas, the drift of fine droplets emitted from the cooling towers contain nearly 6% sodium chloride which deposits on the nearby land areas. This deposition of sodium salts on the nearby agriculture/vegetative lands can convert them into sodic saline or sodic alkaline soils depending on the nature of the soil and enhance the sodicity of ground and surface water. The salt deposition problem from such cooling towers aggravates where national pollution control standards are not imposed or not implemented to minimize the drift emissions from wet cooling towers using seawater make-up.

Respirable suspended particulate matter, of less than 10 micrometers (µm) in size, can be present in the drift from cooling towers. Larger particles above 10 µm in size are generally filtered out in the nose and throat via cilia and mucus but particulate matter smaller than 10 µm, referred to as PM_{10}, can settle in the bronchi and lungs and cause health problems. Similarly, particles smaller than 2.5 µm, ($PM_{2.5}$), tend to penetrate into the gas exchange regions of the lung, and very small particles (less than 100 nanometers) may pass through the lungs to affect other organs. Though the total particulate emissions from wet cooling towers with fresh water make-up is much less, they contain more PM_{10} and $PM_{2.5}$ than the total emissions from wet cooling towers with sea water make-up. This is due to lesser salt content in fresh water drift (below 2,000 ppm) compared to the salt content of sea water drift (60,000 ppm).

Use as a Fuel-gas Stack

Flue gas stack inside a natural draft wet cooling tower.

At some modern power stations equipped with flue gas purification, such as the Großkrotzenburg Power Station and the Rostock Power Station, the cooling tower is also used as a flue-gas stack (industrial chimney), thus saving the cost of a separate chimney structure. At plants without flue gas purification, problems with corrosion may occur, due to reactions of raw flue gas with water to form acids.

Flue gas stack connection into a natural draft wet cooling tower.

Sometimes, natural draft cooling towers are constructed with structural steel in place of concrete (RCC) when the construction time of natural draft cooling tower is exceeding the construction time of the rest of the plant or the local soil is of poor strength to bear the heavy weight of RCC cooling towers or cement prices are higher at a site to opt for cheaper natural draft cooling towers made of structural steel.

Operation in Freezing Weather

Some cooling towers (such as smaller building air conditioning systems) are shut down seasonally, drained, and winterized to prevent freeze damage.

During the winter, other sites continuously operate cooling towers with 4 °C (39 °F) water leaving the tower. Basin heaters, tower draindown, and other freeze protection methods are often employed in cold climates. Operational cooling towers with malfunctions can freeze during very cold weather. Typically, freezing starts at the corners of a cooling tower with a reduced or absent heat load. Severe freezing conditions can create growing volumes of ice, resulting in increased structural loads which can cause structural damage or collapse.

To prevent freezing, the following procedures are used:

- The use of water modulating by-pass systems is not recommended during freezing weather. In such situations, the control flexibility of variable speed motors, two-speed motors, and two-speed motors multi-cell towers should be considered a requirement.
- Do not operate the tower unattended. Remote sensors and alarms may be installed to monitor tower conditions.
- Do not operate the tower without a heat load. Basin heaters may be used to keep the water in the tower pan at an above-freezing temperature. Heat trace ("heating tape") is a resistive heating element that is installed along water pipes to prevent freezing in cold climates.
- Maintain design water flow rate over the tower fill.
- Manipulate or reduce airflow to maintain water temperature above freezing point.

Fire Hazard

Cooling towers constructed in whole or in part of combustible materials can support internal fire propagation. Such fires can become very intense, due to the high surface-volume ratio of the towers, and fires can be further intensified by natural convection or fan-assisted draft. The resulting damage can be sufficiently severe to require the replacement of the entire cell or tower structure. For this reason, some codes and standards recommend that combustible cooling towers be provided with an automatic fire sprinkler system. Fires can propagate internally within the tower structure when the cell is not in operation (such as for maintenance or construction), and even while the tower is in operation, especially those of the induced-draft type, because of the existence of relatively dry areas within the towers.

Structural Stability

Being very large structures, cooling towers are susceptible to wind damage, and several spectacular failures have occurred in the past. At Ferrybridge power station on 1 November 1965, the station was the site of a major structural failure, when three of the cooling towers collapsed owing to vibrations in 85 mph (137 km/h) winds. Although the structures had been built to withstand higher wind speeds, the shape of the cooling towers caused westerly winds to be funneled into the towers themselves, creating a vortex. Three out of the original eight cooling towers were destroyed, and the remaining five were severely damaged. The towers were later rebuilt and all eight cooling towers were strengthened to tolerate adverse weather conditions. Building codes were changed to include improved structural support, and wind tunnel tests were introduced to check tower structures and configuration.

Spent Fuel Pool

Spent fuel pools (SFP) are storage pools for spent fuel from nuclear reactors. They are typically 40 or more feet (12 m) deep, with the bottom 14 feet (4.3 m) equipped with storage racks designed to hold fuel assemblies removed from the reactor. A reactor's pool is specially designed for the reactor in which the fuel was used and situated at the reactor site. An away-from-reactor, Independent Spent Fuel Storage Installation (ISFSI), such as the one located at the Morris Operation, is also sometimes used. In many countries, the fuel assemblies, after being in the reactor for 3 to 6 years, are stored underwater for 10 to 20 years before being sent for reprocessing or dry cask storage. The water cools the fuel and provides shielding from radiation.

While only about 20 feet (about 6 m) of water is needed to keep radiation levels below acceptable levels, the extra depth provides a safety margin and allows fuel assemblies to be manipulated without special shielding to protect the operators.

Operation

About a quarter to a third of the total fuel load of a reactor is removed from the core every 12 to 24 months and replaced with fresh fuel. Spent fuel rods generate intense heat and dangerous radiation that must be contained. Fuel is moved from the reactor and manipulated in the pool generally by automated handling systems, although some manual systems are still in use. The fuel bundles fresh from the core are normally segregated for several months for initial cooling before being sorted into other parts of the pool to wait for final disposal. Metal racks keep the fuel in controlled positions for physical protection and for ease of tracking and rearrangement. High-density racks also incorporate boron-10, often as boron carbide (Metamic, Boraflex, Boral, Tetrabor and Carborundum) or other neutron-absorbing material to ensure subcriticality. Water quality is tightly controlled to prevent the fuel or its cladding from degrading. Current regulations in the United States permit re-arranging of the spent rods so that maximum efficiency of storage can be achieved.

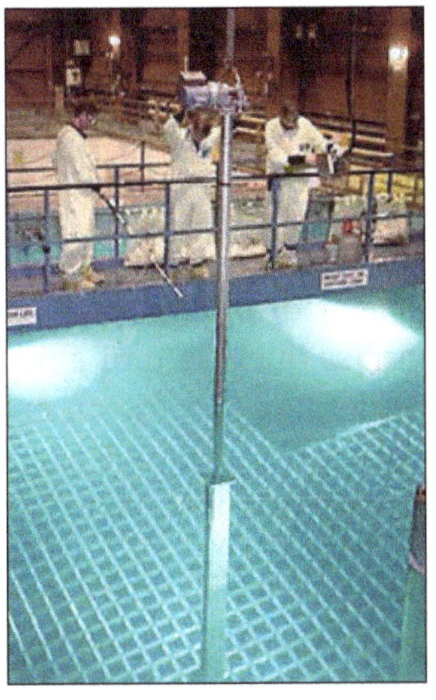

Spent fuel pool.

Worker examines a pond for storing spent fuel rods at thc Lcningrad nuclcar power plant in Sosnovy Bor.

The maximum temperature of the spent fuel bundles decreases significantly between 2 and 4 years, and less from 4 to 6 years. The fuel pool water is continuously cooled to remove the heat produced by the spent fuel assemblies. Pumps circulate water from the spent fuel pool to heat exchangers, then back to the spent fuel pool. The water temperature in normal operating conditions is held below 50 °C (120 °F). Radiolysis, the dissociation of molecules by radiation, is of particular concern in wet storage, as water may be split by residual radiation and hydrogen gas may accumulate increasing the risk of explosions. For this reason the air in the room of the pools, as well as the water, must be continually monitored and treated.

Other Possible Configurations

Rather than manage the pool's inventory to minimize the possibility of continued fission activity, China is building a 200 MWt nuclear reactor to run on used fuel from nuclear power stations to generate process heat for district heating and desalination. Essentially an SFP operated as a deep pool-type reactor; it will operate at atmospheric pressure, which will reduce the engineering requirements for safety.

Other research envisions a similar low-power reactor using spent fuel where instead of limiting the production of hydrogen by radiolysis, it is encouraged by the addition of catalysts and ion scavengers to the cooling water. This hydrogen would then be removed to use as fuel.

Risks

The neutron absorbing materials in spent fuel pools have been observed to degrade severely over time, reducing the safety margins of maintaining subcriticality; in addition, it has been shown that the in-situ measurement technique used to evaluate these neutron absorbers (Boron Areal Density Gauge for Evaluating Racks, or BADGER) has an unknown degree of uncertainty.

If there is a prolonged interruption of cooling due to emergency situations, the water in the spent fuel pools may boil off, possibly resulting in radioactive elements being released into the atmosphere.

In the magnitude 9 earthquake that struck the Fukushima nuclear plants in March 2011, three of the spent fuel pools were in buildings that lost the roof and were seen to be emitting water vapor. The US NRC wrongly stated that the pool at reactor 4 had boiled dry—this was denied at the time by the Japanese and found to be incorrect in subsequent inspection and data examination.

According to nuclear plant safety specialists, the chances of criticality in a spent fuel pool are very small, usually avoided by the dispersal of the fuel assemblies, inclusion of a neutron absorber in the storage racks and overall by the fact that the spent fuel has too low an enrichment level to self-sustain a fission reaction. They also state that if the water covering the spent fuel evaporates, there is no element to enable a chain reaction by moderating neutrons.

According to Dr. Kevin Crowley of the Nuclear and Radiation Studies Board, "successful terrorist attacks on spent fuel pools, though difficult, are possible. If an attack leads to a propagating zirconium cladding fire, it could result in the release of large amounts of radioactive material." After the September 11, 2001 attacks the Nuclear Regulatory Commission required American nuclear plants "to protect with high assurance" against specific threats involving certain numbers and capabilities of assailants. Plants were also required to "enhance the number of security officers" and to improve "access controls to the facilities".

Containment Building

The containment building is a gas-tight building (shell) or other enclosure around a nuclear reactor and a primary circuit. The containment is the most characteristic structure of a nuclear power plant. Practically all nuclear power plants built during the last few decades include a containment building.

The containment building should be designed to ensure or contribute to achievement of the following safety functions:

- Confinement: Confinement of radioactive material in operational states and in accident conditions.
- Protection: Protection of the power plant against external natural and human induced events.
- Shielding: Radiation shielding in operational states and in accident conditions.

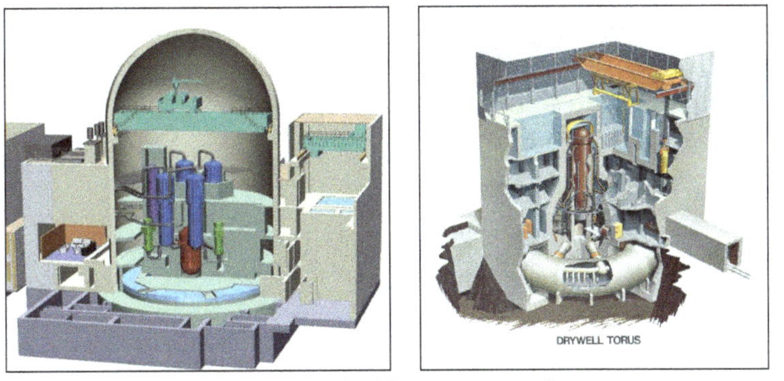

Containment building BWR Mark-I.

Containment vs. Confinement

The International Atomic Energy Agency (IAEA) defines the Confinement as a prevention or control of releases of radioactive material to the environment in operation or in accidents.

Confinement is closely related in meaning to containment, but confinement is typically used to refer to the safety function of preventing the 'escape' of radioactive material, whereas containment refers to the means for achieving that safety function.

The Containment refers to methods or physical structures designed to prevent as low as reasonably achievable the dispersion of radioactive substances.

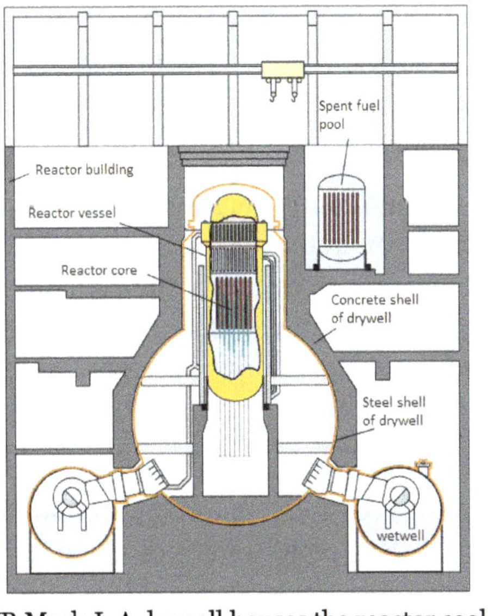

Containment building BWR Mark-I. A drywell houses the reactor coolant system. A wetwell is a suppression chamber, which stores a large body of water.

The containment building is primarily designed to prevent or mitigate the uncontrolled release of radioactive material to the environment in operational states and in accident conditions. Therefore it is considered to be the fourth and final barrier in the Defence in depth strategy.

While the containment plays a crucial role in Design Basis Accidents or in Design Extension conditions, it is "only" designed to condense steam from primary coolant and to contain it inside the building.

In case of Design Basis Accidents such as the Large Break Loss of Coolant Accident (LBLOCA) the pressure increase is usually significant and active containment systems (pressure-suppression systems) must be available in order to maintain the integrity (to keep the pressure and temperature under certain limits) of the containment building.

Containment building BWR Mark-II.

Types of Containment Buildings

- Full pressure dry containment (PWR): Dry containment envelope is a steel shell or a concrete building (cylindrical or spherical) that surrounds the NSSS (Nuclear Steam Supply System). During normal operation a prescribed negative pressure is maintained by a fan cooler system. During accidents the pressure-suppression and the heat sink can be accomplished by a fan cooler system or by a containment water spray system.
- Full pressure double wall containment (PWR): This containment usually consist of a primary containment shell (a steel or concrete shell, basically cylindrical or spherical in shape) and a secondary confinement building (a concrete shell surrounding the primary containment). The principle of the primary containment is similar to that of the full pressure dry containment. The secondary confinement building protects the systems and components against external postulated initiating events and captures leakage from the primary containment.
- Bubbling condenser containment building (PWR): The bubbling condenser containment system uses a concept for the suppression pool or suppression pools in which the high pressure steam following a LOCA or a MSLB conditions is directed through submerged tubes into pools of water. In this pools the steam is condensed and this acts against the pressure increase.
- Ice condenser containment building: The ice condenser containment system uses a concept of ice chambers in which the high pressure steam following a LOCA or a MSLB conditions is directed into chambers containing baskets filled with ice. In this chambers the steam is condensed and this acts against the pressure increase.
- BWR containment buildings: Modern boiling water reactors (BWRs) have no steam generators, so that the reactor coolant system is more compact than that of pressurized water reactors (PWRs). The containment building can be correspondingly smaller. On the other hand BWRs also utilise the large heat sink inside the containment building for energy removal

during reactor isolation events. This is due to the fact the excess of energy is formed by slightly radioactive steam, which cannot be released to the atmosphere. The BWR containments vary widely depending on certain reactor design.The major containment designs are the Mark I, Mark II and the Mark III. The BWR containments consist usually of the following parts:

- Drywell: A drywell houses the reactor coolant system.
- Suppression pool or wetwell: A wetwell is a suppression chamber, which stores a large body of water and therefore it is commonly called as the suppression pool.
- Containment envelope: The Mark III containment has a leak tight, cylindrical, steel containment vessel. This vessel surrounds the drywell and the suppression pool.

The purpose of the drywell and the wetwell (suppression pool) is to reduce the pressure if a LOCA or a MSLB occurs. The steam from a leak in these cases enters the drywell and is directed through submerged tubes into the water of the suppression pool (wetwell), where it condenses, and the pressure in the drywell is reduced.

- Pressurized containments in heavy water reactors. In comparison with LWR containments, most of the HWR containment buildings are accessible, therefore a ventilation is needed for maintaining working conditions. Some units use a single unit suppression, which consist of:
    - A containment envelope comprising a prestressed, post-tensioned concrete reactor building.
    - A large dousing tank, which is elevated around the building dome and a powerful spray system.
    - A long term containment cooling system.

Some units use a multi-unit vacuum building, where more containment envelopes (more units) are connected to a common vacuum building with powerful energy and pressure suppression system.

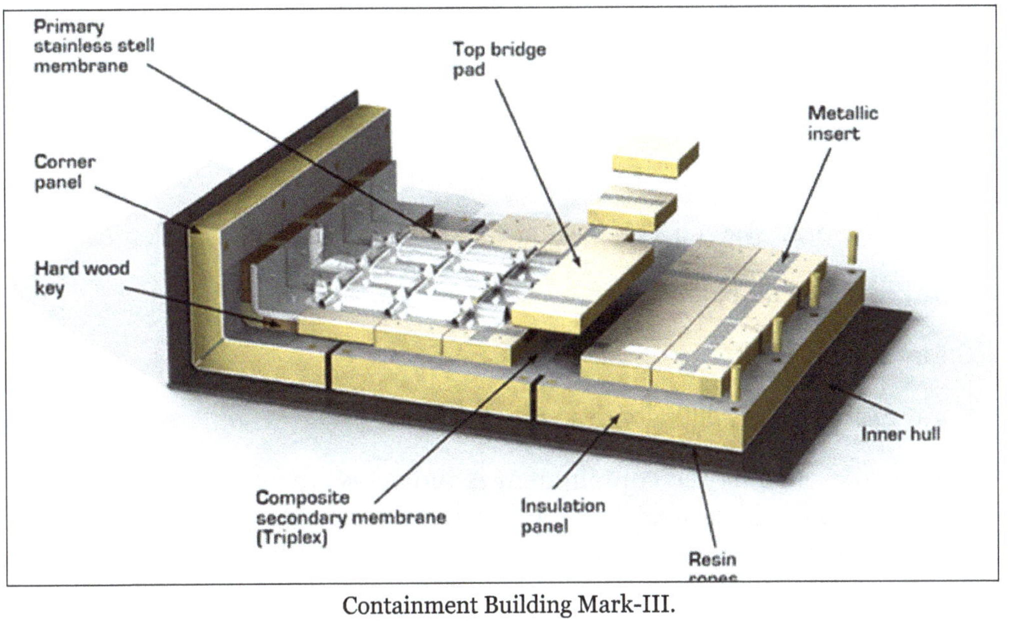

Containment Building Mark-III.

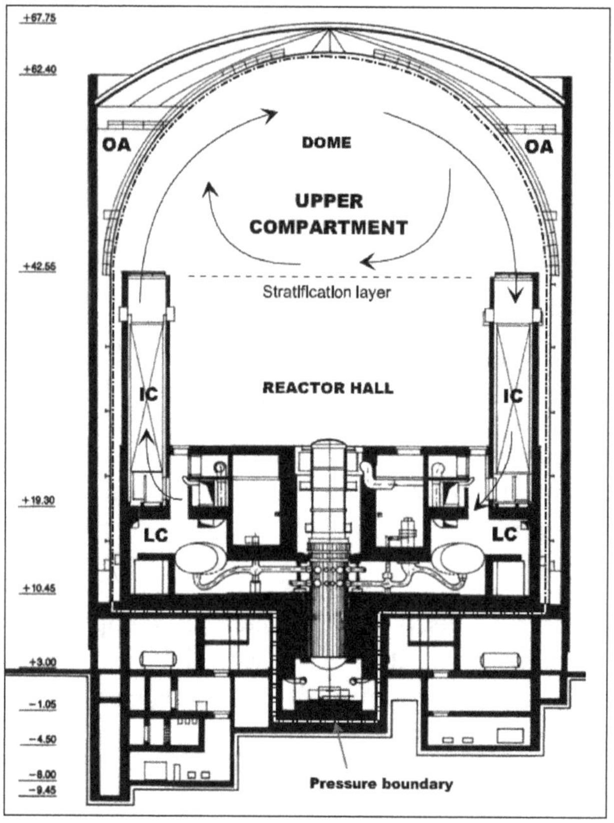

Containment Building – Ice Condenser.

Pressure-Suppression Systems

Pressure-suppression systems are critical to safety and greatly affect the size of containment. Suppression refers to condensing the steam after a major break has released it from the cooling system. There are many designs of suppression systems around the world. Most of Pressurized Water Reactors (PWRs) containments use two stage pressure-suppression systems:

- Fan Cooler System: This system circulates air through heat exchangers and filters to provide the cooling of containment atmosphere. Since this system is not sufficient for suppression during severe loss of coolant accidents, the containment spray system must be available as the secondary active pressure-suppression system.
- Containment Spray System: This system consist usually from three elements:
  - Spray System Pump,
  - Spray System Tank,
  - Spray System Rings and Nozzles.

When pressure increase inside the containment is indicated, the containment spray system is automatically started and the pumps (usually with 3×100% redundancy) take a suction from the tank (refueling water storage tank can be also used) and pump the water into spray nozzles located in the upper part of the containment. The water droplets, being cooler than the steam, will remove heat from the steam, which will cause the steam to condense. This will cause a reduction in the pressure of the building and will also reduce the temperature of the containment atmosphere. The spray system usually contains extra chemical additives dissolved in the tank to enhance the removal of particular radionuclides from the containment atmosphere. Especially radioiodine, which is of particular importance, can be effectively bonded to potassium hydroxide or sodium hydroxide.

Most of Boiling Water Reactors (BWR) containments use pressure-suppression pools in order to maintain the integrity of the containment building. The major containment designs are the Mark I, Mark II and the Mark III. The Mark I and Mark II containments consist of two main parts:

- Drywell: A drywell houses the reactor coolant system.
- Wetwell: A wetwell is a suppression chamber, which stores a large body of water and therefore it is commonly called as the suppression pool.

Water spray systems are usually installed in both the drywell and the wetwell. The Mark III design consists of a primary containment and a drywell.

Containment buildings and containment pressure-suppression systems vary widely depending on certain reactor design. In some cases really unique technologies can be installed. For example, the containment building of Loviisa NPP uses two ice condensers as the pressure-suppression system.

Ice Condenser Containment Building

The ice condenser is a safety system designed to rapidly absorb steam and reduce containment pressure following a Loss of Coolant Accident (LOCA) or Main Steam Line Break (MSLB). The high pressure steam following LOCA or MSLB conditions is directed into chambers containing baskets filled with ice. Ice blocks, being naturally cooler than the steam, will remove heat from the steam, which will cause the steam to condense. This will cause a reduction in the pressure of the building and will also reduce the temperature of the containment atmosphere.

This system has many positive features. Pressure suppression via ice condensers is very effective because of the ability of ice to absorb high amount of energy during a phase transition. This feature allows containment building to be much smaller than conventional containment buildings. Moreover water from melting of ice can be used for passive cooling and flooding of the small-sized cavity (in the case of Loviisa NPP), where the reactor is located.

Ice condenser containments were designed by Westinghouse in the late 1970s and early 1980s, therefore in the United States there are 9 ice condenser containments on 5 sites. Except Loviisa NPP there is one other ice condenser unit in the world, the Ohi plant in Japan.

Hydrogen Mitigation in Water Cooled Power Reactors

Hydrogen mitigation in water cooled power reactors is an important area of study in the field of safety of nuclear reactors. Hydrogen and oxygen can be generated during normal operation of a power reactor primarily as a consequence of the radiolysis of the water in the core.

During accidents hydrogen and oxygen can be generated also as a consequence of:

- Metal–water reactions in the core.
- Radiolysis of the water in the sump or the suppression pool.
- Degassing of hydrogen dissolved in the primary coolant.
- Chemical reactions with materials in the containment (interactions of molten core debris with concrete).
- Thermolysis of water – in extreme cases.

During DBAs (Design Basis Accidents) such as the large break loss of coolant accident, the production of hydrogen (metal–water reactions in the core) is limited at low values by the operation of the emergency core cooling systems. Also the radiolysis of the water in the core is relatively slow process. Therefore, from the DBAs point of view the hydrogen hazards can be eliminated by maintaining the local hydrogen concentration below its flammability limit (4% of volume). This requirement can be ensured by mixing devices or thermal hydrogen recombiners.

In the past two decades, research activities to analyse the threat of hydrogen in a post-accident condition have focused mainly on the mitigation of hydrogen hazards in severe accidents. In severe accidents consider the possibility of the large scale core degradation and even the possibility of molten core concrete interactions.

For these cases the capacity of conventional DBA hydrogen control is insufficient. During severe accidents, local hydrogen concentrations can exceed its flammability limit in a short time. analysis of the hydrogen threat in post-accident containments is complex and highly plant- and scenario-specific.

In order to avoid the threat of hydrogen explosion, mitigation measures for hydrogen in severe accidents must be implemented. The strategy is usually based on following physical principles:

- Avoiding of formation of flammable mixtures by oxygen control.
- Avoiding of formation of flammable mixtures by hydrogen control.
- Avoiding of flammable or detonable concentrations of hydrogen by controlled ignition.

The most effective measures are:

- Passive or active air mixing systems.
- Post accident dilution (PAD) of containment atmosphere by inert gas injection.
- Catalytic recombination of hydrogen.
- Controlled ignition by deliberated ignition system (DIS).

References

- Nuclear-power-plant, encyclopedia: energyeducation.ca, Retrieved 5 January, 2019

- "Radioactive Waste Management". World Nuclear Association. Archived from the original on 21 September 2013. Retrieved 19 September 2013
- Nuclear-reactor, technology: britannica.com, Retrieved 6 February, 2019
- Van Vliet, Michelle T. H.; Wiberg, David; Leduc, Sylvain; Riahi, Keywan (4 January 2016). "Power-generation system vulnerability and adaptation to changes in climate and water resources". Nature Climate Change. 6 (4): 375–380. Doi:10.1038/nclimate2903. Retrieved 28 March 2016
- Reactor-coolant-pump: nuclear-power.net, Retrieved 7 March, 2019
- Ryan K.J.; Ray C.G. (editors) (2004). Sherris Medical Microbiology (4th ed.). McGraw Hill. ISBN 978-0-8385-8529-0
- Containment-building, nuclear-power-plant: nuclear-power.net, Retrieved 8 April, 2019
- Cleanenergy Footprints (cleanenergy.org). Identifying Nuclear Reactors in Google Earth Retrieved 5/19/2014

Permissions

All chapters in this book are published with permission under the Creative Commons Attribution Share Alike License or equivalent. Every chapter published in this book has been scrutinized by our experts. Their significance has been extensively debated. The topics covered herein carry significant information for a comprehensive understanding. They may even be implemented as practical applications or may be referred to as a beginning point for further studies.

We would like to thank the editorial team for lending their expertise to make the book truly unique. They have played a crucial role in the development of this book. Without their invaluable contributions this book wouldn't have been possible. They have made vital efforts to compile up to date information on the varied aspects of this subject to make this book a valuable addition to the collection of many professionals and students.

This book was conceptualized with the vision of imparting up-to-date and integrated information in this field. To ensure the same, a matchless editorial board was set up. Every individual on the board went through rigorous rounds of assessment to prove their worth. After which they invested a large part of their time researching and compiling the most relevant data for our readers.

The editorial board has been involved in producing this book since its inception. They have spent rigorous hours researching and exploring the diverse topics which have resulted in the successful publishing of this book. They have passed on their knowledge of decades through this book. To expedite this challenging task, the publisher supported the team at every step. A small team of assistant editors was also appointed to further simplify the editing procedure and attain best results for the readers.

Apart from the editorial board, the designing team has also invested a significant amount of their time in understanding the subject and creating the most relevant covers. They scrutinized every image to scout for the most suitable representation of the subject and create an appropriate cover for the book.

The publishing team has been an ardent support to the editorial, designing and production team. Their endless efforts to recruit the best for this project, has resulted in the accomplishment of this book. They are a veteran in the field of academics and their pool of knowledge is as vast as their experience in printing. Their expertise and guidance has proved useful at every step. Their uncompromising quality standards have made this book an exceptional effort. Their encouragement from time to time has been an inspiration for everyone.

The publisher and the editorial board hope that this book will prove to be a valuable piece of knowledge for students, practitioners and scholars across the globe.

Index

www.ingramcontent.com/pod-product-compliance
Lightning Source LLC
LaVergne TN
LVHW080719210926
848485LV00003B/491